United States Government Accountability Office

Report to Congressional Committees

I0488954

September 2012

ORGANIZATIONAL TRANSFORMATION

Enterprise Architecture Value Needs to Be Measured and Reported

GAO

Accountability ★ Integrity ★ Reliability

GAO-12-791

Highlights of GAO-12-791, a report to congressional committees

ORGANIZATIONAL TRANSFORMATION

Enterprise Architecture Value Needs to Be Measured and Reported

Why GAO Did This Study

According to OMB, the federal executive branch plans to spend at least $75 billion on information technology (IT) investments in fiscal year 2012. In response to a statute which mandates that GAO identify duplicative activities within federal agencies, GAO previously identified enterprise architecture as a mechanism for reducing duplication and overlap in investments. An architecture is a "blueprint" that describes how an organization operates in terms of business processes and technology, how it intends to operate in the future, and how it plans to transition to the future state. Knowing whether architecture outcomes are being achieved requires defining the architecture's goals, establishing a method and metrics to measure architecture outcomes, and periodically measuring and reporting these outcomes. To assess agencies' use of architecture as a mechanism for reducing duplication and overlap, GAO committed to determine the extent to which agencies are measuring and reporting architecture outcomes and benefits. To do this, GAO reviewed relevant documentation from 27 major federal agencies, reviewed the results of a GAO survey on the benefits of using architecture, and interviewed agency officials.

What GAO Recommends

GAO is making recommendations to the agencies and OMB to improve measurement and reporting of architecture outcomes. In commenting on a draft of this report, OMB and most of the agencies generally agreed with the findings and recommendations.

View GAO-12-791. For more information, contact Valerie C. Melvin at (202) 512-6304 or melvinv@gao.gov.

What GAO Found

Among the 27 agencies that GAO studied, all have fully or partially defined goals or purposes for their architectures, 11 have fully or partially established a method or metrics for measuring outcomes resulting from the use of their architectures, while 5 have fully or partially measured and reported outcomes and benefits (see table).

Agency	Goals or purpose defined	Metrics and method established	Outcomes and benefits periodically measured and reported
Agriculture	●	○	○
Air Force	●	○	○
Army	●	◑	○
Commerce	●	◑	○
Defense—Business Enterprise Architecture	●	○	○
Defense—Enterprise Architecture	●	○	○
Education	●	◑	◑
Energy	●	○	○
Health and Human Services	●	●	◑
Homeland Security	●	○	○
Housing and Urban Development	●	●	◑
Interior	●	○	○
Justice	●	○	○
Labor	●	○	○
Navy	●	○	○
State	●	○	○
Transportation	●	◑	○
Treasury	●	◑	◑
Veterans Affairs	◑	○	○
Environmental Protection Agency	●	○	○
General Services Administration	●	◑	○
National Aeronautics and Space Administration	●	○	○
National Science Foundation	●	○	○
Nuclear Regulatory Commission	●	◑	○
Officer of Personnel Management	●	◑	○
Small Business Administration	●	○	○
Social Security Administration	●	○	○
United States Agency for International Development	●	●	●

Source: GAO analysis of agency data.

Agencies cited a lack of guidance as a key reason why they have not established methods and metrics for measuring outcomes and benefits. Although the Office of Management and Budget (OMB) has issued recent enterprise architecture guidance to agencies, OMB has not yet provided sufficient details on the method and metrics that could be used to measure architecture program outcomes.

_____ **United States Government Accountability Office**

Contents

Tables

Abbreviations

CIO	chief information officer
Commerce	Department of Commerce
DHS	Department of Homeland Security
DOD	Department of Defense
EAMMF	Enterprise Architecture Management Maturity Framework
Education	Department of Education
Energy	Department of Energy
EPA	Environmental Protection Agency
GSA	General Services Administration
HHS	Department of Health and Human Services
HUD	Department of Housing and Urban Development
Interior	Department of the Interior
IT	information technology
Justice	Department of Justice
Labor	Department of Labor
NASA	National Aeronautics and Space Administration
NRC	Nuclear Regulatory Commission
NSF	National Science Foundation
OMB	Office of Management and Budget
OPM	Office of Personnel Management
SBA	Small Business Administration
SSA	Social Security Administration
State	Department of State
Transportation	Department of Transportation
Treasury	Department of the Treasury
USAID	United States Agency for International Development
USDA	United States Department of Agriculture
VA	Department of Veterans Affairs

United States Government Accountability Office
Washington, DC 20548

September 26, 2012

The Honorable Joseph Lieberman
Chairman
The Honorable Susan Collins
Ranking Member
Committee on Homeland Security and Governmental Affairs
United States Senate

The Honorable Darrell Issa
Chairman
The Honorable Elijah Cummings
Ranking Member
Committee on Oversight and Government Reform
House of Representatives

Billions of taxpayer dollars are spent on information technology (IT) investments each year; according to the Office of Management and Budget (OMB), the executive branch plans to spend at least $75 billion in fiscal year 2012. We have previously reported that federal expenditures on IT could be reduced by, among other things, using enterprise architecture as a tool for organizational transformation.[1]

An enterprise architecture is a blueprint for organizational change defined in models that describe (in both business and technology terms) how the entity operates today and how it intends to operate in the future; it also includes a plan for transitioning to this future state. Effective use of an enterprise architecture is a hallmark of successful organizations and can be important to achieving operations and technology environments that maximize institutional mission performance and outcomes. Among other things, this includes realizing cost savings through consolidation and reuse of shared services and elimination of antiquated and redundant mission operations, enhancing information sharing through data standardization and system integration, and optimizing service delivery through streamlining and normalization of business processes and

---

[1]GAO, *Opportunities to Reduce Potential Duplication in Government Programs, Save Tax Dollars, and Enhance Revenue*, GAO-11-318SP (Washington, D.C.: Mar. 1, 2011). An interactive, web-based version of the report is available at http://www.gao.gov/ereport/gao-11-318SP.

GAO-12-791 Organizational Transformation

mission operations. Moreover, the use of architectures is required by the Clinger-Cohen Act of 1996 and by OMB.[2]

In our March 2011 report on opportunities to reduce potential duplication in government programs,[3] we identified enterprise architecture as a mechanism for identifying potential overlap and duplication. We noted that realizing this potential and knowing whether benefits are in fact being achieved from the use of an architecture requires associated measures and metrics. Accordingly, under the statutory requirement which mandates that GAO identify federal programs, agencies, offices, and initiatives with duplicative goals and activities within departments and government-wide,[4] we committed to study the extent to which federal departments and agencies are measuring and reporting enterprise architecture outcomes and benefits.

To address our objective, we analyzed 27 major departments' and agencies'[5] documents describing their enterprise architecture goals and purposes and their approaches to measuring and reporting architecture outcomes and benefits, such as their IT Strategic Plan, Enterprise Architecture Program Management Plan, and Enterprise Architecture Value Measurement Plan. We compared the agencies' approaches to relevant elements of our Enterprise Architecture Management Maturity Framework (EAMMF).[6] Further, we reviewed outcomes reported to agency enterprise architecture oversight officials and analyzed responses to a 2011 GAO survey about the benefits associated with agencies' architecture programs. We also discussed our analyses with and obtained testimonial evidence from cognizant agency officials. A more detailed

---

[2]40 U.S.C. § 11315; The E-Government Act of 2002 also provided a more detailed definition of the concept and elements of enterprise architecture. See 44 U.S.C. §§ 3601(4) and 3602; OMB Circular A-130 (Nov. 30, 2000); and Chief Information Officers Council, *A Practical Guide to Federal Enterprise Architecture*, Version 1.0 (February 2001).

[3]GAO-11-318SP.

[4]Pub. L. No. 111-139, § 21, 124 Stat. 29 (2010), 31 U.S.C. § 712 Note.

[5]These 27 major departments and agencies are the 24 Chief Financial Officer Act entities identified in 31 U.S.C. § 901(b), as well as the Departments of the Air Force, Army, and Navy.

[6]GAO, *Organizational Transformation: A Framework for Assessing and Improving Enterprise Architecture Management (Version 2.0)*, GAO-10-846G (Washington, D.C.: August 2010).

discussion of our objective, scope, and methodology is provided in appendix I.

We conducted this performance audit from November 2011 to September 2012 in accordance with generally accepted government auditing standards. Those standards require that we plan and perform the audit to obtain sufficient, appropriate evidence to provide a reasonable basis for our findings and conclusions based on our audit objective. We believe that the evidence obtained provides a reasonable basis for our findings and conclusions based on our audit objective.

Background

An enterprise architecture is a blueprint that describes the current and desired states of an organization or functional area in both logical and technical terms, as well as a plan for transitioning between the two states. An enterprise can be viewed as either a single organization or a functional area that transcends more than one organization. An architecture can be viewed as the structure (or structural description) of any activity. Thus, enterprise architectures are systematically derived and captured descriptions depicted in models, diagrams, and narratives. More specifically, an architecture describes the enterprise in logical terms (such as interrelated business processes and business rules, information needs and flows, and work locations and users) as well as in technical terms (such as hardware, software, data, communications, security attributes, and performance standards). It provides these perspectives both for the enterprise's current environment and for its target environment, and it provides a transition plan for moving from the current to the target environment. Enterprise architectures are a recognized tenet of organizational transformation and IT management in public and private organizations.

When employed in concert with other institutional management disciplines, such as strategic planning, portfolio-based capital planning and investment control, and human capital management, an enterprise architecture can greatly increase the chances of configuring an organization to promote agility and responsiveness, optimize mission performance and strategic outcomes, and address new federal initiatives like promoting open and participatory government and leveraging cloud computing.

Federal Legislation and OMB Guidance Pertaining to Establishment of an Enterprise Architecture

The Clinger-Cohen Act of 1996, among other things, requires federal agency chief information officers (CIO) to develop, maintain, and facilitate the implementation of IT architectures.[7] Subsequent OMB guidance more broadly interpreted IT architecture as an enterprise architecture.[8] In September 1999, the federal CIO Council published the Federal Enterprise Architecture Framework,[9] which provided federal agencies with a common construct for their architectures to facilitate the coordination of common business processes, technology insertion, information flows, and system investments among federal agencies. The framework defined a collection of interrelated models for describing multiorganizational functional segments of the federal government.[10] Further, in 2000 and 2001, the federal CIO Council developed enterprise architecture guidance focused on assessing an IT investment's compliance with an architecture[11] as well as guidance that addressed the end-to-end steps associated with developing, maintaining, and implementing an architecture program.[12]

OMB is responsible for overseeing the development of enterprise architectures within and across federal agencies.[13] In February 2002, it

---

[7]40 U.S.C. § 11315. According to GAO's EAMMF, such architectures provide an important means of integrating business processes and agency goals with IT.

[8]See for example OMB, *Information Technology Architectures*, Memorandum M-97-16 (June 18, 1997), rescinded with the update of OMB Circular A-130 (Nov. 30, 2000), which requires that agencies document and submit their enterprise architecture to OMB. Chief Information Officers Council, *Architecture Alignment and Assessment Guide* (October 2000). Chief Information Officers Council, *A Practical Guide to Federal Enterprise Architecture*, Version 1.0 (February 2001).

[9]Federal Enterprise Architecture Framework, Version 1.1 (September 1999).

[10]The most recent revision to the Federal Enterprise Architecture Framework (Version 2.0) is included in OMB's Common Approach to Federal Enterprise Architecture, which is discussed subsequently in this report.

[11]Chief Information Officers Council, *Architecture Alignment and Assessment Guide* (October 2000).

[12]Chief Information Officers Council, *A Practical Guide to Federal Enterprise Architecture*, Version 1.0 (February 2001).

[13]The E-Government Act of 2002 provided a more detailed definition of the concept and elements of enterprise architecture and established the OMB Office of Electronic Government and assigned it, among other things, responsibilities for overseeing the development of enterprise architectures within and across federal agencies. See 44 U.S.C. § 3601(4) and 44 U.S.C § 3602(f)(14).

established the Federal Enterprise Architecture program. According to OMB, the program is intended to facilitate government-wide improvement through cross-agency analysis and identification of duplicative investments, gaps, and opportunities for collaboration, interoperability, and integration within and across agency programs. Federal enterprise architecture reference models are intended to inform agency efforts to develop their agency-specific enterprise architectures and enable agencies to ensure that their proposed investments are not duplicative with those of other agencies and to pursue, where appropriate, joint projects. In 2007, OMB issued the Federal Enterprise Architecture Practice Guidance[14] to provide high-level overviews of architecture concepts, descriptions of the content included in architecture work products, and direction on developing and using architectures, including measuring enterprise architecture program value.

According to the latest version of OMB's Enterprise Architecture Assessment Framework (version 3.1, dated June 2009),[15] its purpose is to provide the measurement areas and criteria for federal agencies to use in realizing architecture-driven performance improvements and outcomes (e.g., improving mission performance; saving money and avoiding costs; enhancing the quality of agency investment portfolios; improving the quality, availability, and sharing of data and information; and increasing the transparency of government operations). To accomplish this, the framework uses key performance indicators to assess architecture maturity or effectiveness relative to three capability areas—completion, use, and results. Each capability area contains a set of key performance indicators and associated outcomes, as well as criteria for gauging progress in meeting the outcomes. In particular, according to the framework, as part of the results capability area, agencies should measure actual results attributed to the architecture, and therefore the effectiveness and value of architecture activities. However, to reduce the reporting burden on agencies, in August 2009, OMB issued a memorandum that stated that agencies were no longer required to provide self-assessments of enterprise architecture completion, use, and results to OMB.

---

[14]OMB, *Federal Enterprise Architecture Practice Guidance* (November 2007).

[15]OMB, *Improving Agency Performance Using Information and Information Technology* (Enterprise Architecture Assessment Framework v3.1) (June 2009).

GAO-12-791 Organizational Transformation

In May 2012, OMB released the *Common Approach to Federal Enterprise Architecture*[16] to promote increased levels of mission effectiveness by standardizing the development and use of architectures within and between federal agencies. The approach stresses that enterprise architecture can enable service delivery, functional integration, and resource optimization, and can be an authoritative reference for the design and documentation of systems and services. According to a memorandum accompanying the *Common Approach*, each agency is to submit to OMB by August 31, 2012, an enterprise roadmap that covers fiscal years 2012 to 2015, to serve as an authoritative reference for IT portfolio reviews.[17] The roadmap is to map the organization's strategic goals to business services and integrate technology solutions across the agency's lines of business. It is to discuss the overall architecture and identify performance gaps, resource requirements, planned solutions, transition plans, and a summary of the current and future architectures. It is also to describe the enterprise architecture governance process, implementation methodology, and documentation framework.

As one of the elements intended to ensure that agency enterprise architecture programs can be effective in developing solutions that support planning and decision making, the guidance begins to lay out a Collaborative Planning Methodology. The methodology entails defining what benefits will be achieved, when those benefits will be achieved, and how those benefits will be measured, as well as measuring performance outcomes against identified metrics. The guidance emphasizes the importance of measuring the attainment of outcomes, so that the positive effects (added value) of the architecture program can be identified. Specifically, each agency's roadmap is to document how the effectiveness and efficiency of the program will be measured. The guidance discusses the difference between outcome and output measures, and notes that while output measures are important for indicating an initiative's progress, outcome measures are needed to indicate the attainment of goals. According to the Federal Chief

---

[16]OMB, *The Common Approach To Federal Enterprise Architecture* (May 2012).

[17]OMB, Memorandum for Federal Agency Chief Information Officers, *Increasing Shared Approaches to Information Technology Services* (Washington, D.C., May 2, 2012). Agencies will be required to submit an updated enterprise roadmap to OMB by April 1st each year, beginning April 1, 2013.

GAO-12-791 Organizational Transformation

Enterprise Architect, OMB plans to provide agencies more detailed guidance on measuring enterprise architecture value by December 2012.

To assist in developing this guidance, the Architecture Subcommittee of the CIO Council's Strategy and Planning Committee has established a working group to develop an approach for measuring enterprise architecture value through identifying best practices from the public and private sectors.[18] The group plans to draw upon research to create a value measurement program that aligns with the Collaborative Planning Methodology discussed in the *Common Approach*, and deliver a white paper to OMB on value measurement by the end of fiscal year 2012. According to the Federal Chief Enterprise Architect, the working group's recommendations will be considered for incorporation into the *Common Approach*.

GAO's Enterprise Architecture Management Maturity Framework

In August 2010, we issued an Enterprise Architecture Management Maturity Framework that provides federal agencies with a common benchmarking tool for assessing the management of their enterprise architecture efforts and developing improvement plans.[19] The framework includes 59 core elements, or building blocks, of enterprise architecture management. The core elements represent practices, structures, activities, and conditions that, when properly employed based on the unique facts and circumstances of each organization and the stated purpose of its enterprise architecture program, can permit that organization to maximize its chances of realizing an architecture's institutional value. The core elements are categorized into seven hierarchical stages of management maturity and four critical success attribute representations.

In particular, core element 41 describes the practice of measuring and reporting enterprise architecture outcomes. The architecture is a strategic asset that represents an investment in the organization's future and is intended to produce strategic mission value (results and outcomes). Measuring the extent to which this expected value is actually being realized is important to identifying what, if any, enterprise architecture

---

[18]The CIO Council includes CIOs and Deputy CIOs from 28 federal agencies and is chaired by the Office of Management and Budget Deputy Director for Management.

[19]GAO-10-846G.

program changes are warranted. Such value can be derived from realizing cost savings through consolidation and reuse of shared services and elimination of antiquated and redundant mission operations, enhancing information sharing through data standardization and system integration, and optimizing service delivery through streamlining and normalization of business processes and mission operations.

In addition, core element 58 specifies that enterprise architecture quality- and outcomes-measurement methods should be continuously improved. Organizations should periodically reevaluate their methods for assessing corporate and subordinate architecture quality and program outcomes and address the extent to which program measures and metrics are sufficiently measurable, meaningful, repeatable, consistent, actionable, and aligned with the architecture program's strategic goals and intended purpose.

Prior GAO Work Has Highlighted Federal Agency Enterprise Architecture Challenges

In 2002 and 2003, we reported on the status of enterprise architectures government-wide.[20] We found that some federal agencies had begun to establish the management foundation needed to successfully develop, implement, and maintain an enterprise architecture, but that executive leadership was key to addressing management challenges identified by enterprise architecture programs: (1) overcoming limited executive understanding, (2) inadequate funding, (3) insufficient number of skilled staff, and (4) organizational parochialism. Accordingly, we made recommendations to OMB to improve enterprise architecture leadership and oversight. OMB responded to these recommendations by establishing its Chief Architects Forum to, among other things, share enterprise architecture best practices among federal agencies, and by developing an assessment tool, which it used to periodically evaluate enterprise architecture programs at federal agencies.

---

[20]GAO, *Information Technology: Enterprise Architecture Use across the Federal Government Can Be Improved*, GAO-02-6 (Washington, D.C.: Feb. 19, 2002); *Information Technology: Leadership Remains Key to Agencies Making Progress on Enterprise Architecture Efforts*, GAO-04-40 (Washington, D.C.: Nov. 17, 2003).

In 2006, we reviewed enterprise architecture management at 27 major federal departments and agencies.[21] Our work showed that the state of architecture development and implementation varied considerably across departments and agencies, with some having more mature programs than others. Overall, most agencies had not reached a sufficient level of maturity in their enterprise architecture development, particularly with regard to their approaches to assessing each investment's alignment with the architecture and measuring and reporting on architecture results and outcomes. Our 2006 report also noted that challenges we identified in our earlier reviews continued to present hurdles to effective implementation of enterprise architecture.

We have also reported on enterprise architecture management and development at several individual departments and agencies, including agencies that have demonstrated improvements to their architectures:

- In 2009, we reported that recent versions of the Department of Homeland Security's (DHS) enterprise architecture had largely addressed our prior recommendations aimed at adding needed architectural depth and breadth.[22] Nonetheless, we concluded that important content, such as prioritized segments and information exchanges between critical business processes, was still missing from its architecture.

- Between 2009 and 2012, we conducted several reviews of the Department of Housing and Urban Development's (HUD) enterprise architecture and made a number of recommendations for

---

[21]GAO, *Enterprise Architecture: Leadership Remains Key to Establishing and Leveraging Architectures for Organizational Transformation*, GAO-06-831 (Washington, D.C.: Aug. 14, 2006).

[22]GAO, *Homeland Security: Despite Progress, DHS Continues to Be Challenged in Managing Its Multi-Billion Dollar Annual Investment in Large-Scale Information Technology Systems*, GAO-09-1002T (Washington, D.C.: Sept. 15, 2009).

improvement.[23] Over the course of these reviews, we found that HUD had made progress in establishing its architecture, although as of September 2012, the department had not yet finalized its updated architecture policy, as we had recommended.

- In September 2011, we reported on the status of the three military departments' (Air Force, Army, and Navy) architecture programs.[24] We reported that while each of the military departments had long-standing efforts to develop and use enterprise architectures, they had much to do before their efforts could be considered mature. Accordingly, we recommended that the military departments each develop a plan for fully satisfying the elements of our framework. The Department of Defense (DOD) and the Army concurred with these recommendations, but the Air Force and Navy did not. In this regard, DOD stated that the Air Force and Navy did not have a valid business case that would justify the implementation of all the elements. However, we maintained that the recommendation was warranted. To date, none of the military departments have addressed our recommendations.

- In April 2012, we reported that the Social Security Administration (SSA) had developed an enterprise architecture for years 2011 through 2016 that captured certain foundational information about the current and target environments to assist in evolving existing information systems and developing new systems; however, the architecture lacked important content that would allow the agency to more effectively plan its investments to reach its vision of modernized systems and operations.[25] We recommended that SSA develop an

---

[23]GAO, *Information Technology: HUD Needs to Strengthen Its Capacity to Manage and Modernize Its Environment*, GAO-09-675 (Washington, D.C.: July 31, 2009); *Information Technology: HUD Needs to Better Define Commitments and Disclose Risk for Modernization Projects in Future Expenditure Plans*, GAO-11-72 (Washington, D.C.: Nov. 23, 2010); *Information Technology: HUD's Expenditure Plan Satisfies Statutory Conditions, and Implementation of Management Controls Is Under Way*, GAO-11-762 (Washington, D.C.: Sept. 7, 2011); and *HUD Information Technology: More Work Remains to Implement Necessary Management Controls*, GAO-12-580T (Washington, D.C.: Mar. 29, 2012).

[24]GAO, *Organizational Transformation: Military Departments Can Improve their Enterprise Architecture Programs*, GAO-11-902 (Washington, D.C.: Sept. 26, 2011).

[25]GAO, *Social Security Administration: Improved Planning and Performance Measures Are Needed to Help Ensure Successful Technology Modernization*, GAO-12-495 (Washington, D.C: Apr. 26, 2012).

enterprise architecture plan that included certain key elements. The agency responded that it would comply with recent direction from the Federal Chief Architect to deliver an enterprise architecture roadmap that meets OMB standards.

In addition to our evaluation of agency-specific enterprise architectures, we have reported on the need for federal agencies to measure and report architecture outcomes. Specifically, in March 2011, we reported that while some progress had been made in improving the content and use of departments' and agencies' architectures, more time was needed for agencies to fully realize the value of having well-defined and implemented architectures.[26] We noted that some agencies had reported that they were addressing the EAMMF core element associated with measuring and reporting enterprise architecture results and outcomes and had realized significant financial benefits. For example, we reported that the Department of the Interior had demonstrated that it was using its enterprise architecture to modernize agency IT operations and avoid costs through enterprise software license agreements and hardware procurement consolidation, which had resulted in reported financial benefits of at least $80 million. However, we concluded that over 50 percent of the departments and agencies had yet to fully address this element.

In February 2012, we again reviewed the extent to which major federal agencies had reported financial benefits from the use of enterprise architecture. We found that four agencies (in addition to the Department of Interior) had done so. These four agencies were the Department of Health and Human Services (HHS), which, facilitated by its architecture program, moved to a new telecommunications contract, resulting in a savings of about $21 million; the Nuclear Regulatory Commission (NRC), which avoided an estimated $1.3 million cost in 2011 by eliminating duplicative staff planning systems; DOD, which reported saving $179 million between fiscal years 2008 and 2010 by streamlining Navy business operations, retiring legacy systems, and moving toward a real-time paperless business environment for processing vendor payments; and the Department of Agriculture, which reported savings of $27 million over 5 years (2011 through 2015) by moving 120,000 e-mail users to a cloud-based solution.[27] We also noted that 12 agencies had reported

---

[26]GAO-11-318SP.

[27]Cloud computing is a form of computing that relies on Internet-based services and resources to provide computing services to customers.

financial benefits but had not reliably measured them (i.e., they did not provide supporting documentation), and an additional 10 agencies had not reported financial benefits, although 8 of these agencies reported that they had established or expected to establish a process to measure benefits in the future.[28]

Almost All Agencies Had Defined the Purpose of Their Architectures, but Had Yet to Fully Measure and Report Outcomes and Benefits

Most of the 27 major agencies in our current study had yet to periodically (i.e., regularly and repeatedly, such as monthly, quarterly or annually) measure and report enterprise architecture outcomes and benefits. Our framework[29] recognizes that knowing whether architecture outcomes are being achieved requires an approach to measuring the value of architecture activities that includes defining the architecture's intended purpose or strategic goals; establishing metrics along with a method to measure architecture outcomes and benefits; and periodically measuring and reporting to the agency's architecture executive committee (executive-level representatives from each line of business, who have the authority to commit resources) these outcomes and benefits. While all agencies had fully or partially defined their architecture's strategic goals or intended purpose, only 3 had fully and 8 had partially established metrics and a method to measure outcomes and benefits. Of the agencies that fully or partially established a method and metrics, 4 had measured and reported outcomes only once, and 1 had periodically (e.g., monthly) reported on outcomes and benefits. A summary of the 27 agencies' progress in measuring and reporting architecture outcomes and benefits is presented in table 1. For detailed assessments of individual departments and agencies against relevant elements of our framework, see appendix II.

---

[28]GAO, *Follow-up on 2011 Report: Status of Actions Taken to Reduce Duplication, Overlap, and Fragmentation, Save Tax Dollars, and Enhance Revenue*, GAO-12-453SP (Washington, D.C.: Feb. 28, 2012).

[29]GAO-10-846G.

Agency	The enterprise architecture's strategic goals or intended purpose are defined	Metrics and a method have been established to measure enterprise architecture strategic mission value (outcomes and benefits)	Enterprise architecture outcomes and benefits are periodically measured and reported to the architecture executive committee
Department of Agriculture	●	○	○
Department of the Air Force	●	○	○
Department of the Army	●	◑	○
Department of Commerce	●	◑	○
Department of Defense – Business Enterprise Architecture	●	○	○
Department of Defense - Enterprise Architecture[a]	●	○	○
Department of Education	●	◑	◑
Department of Energy	●	○	○
Department of Health and Human Services	●	●	◑
Department of Homeland Security	●	○	○
Department of Housing and Urban Development	●	●	◑
Department of the Interior	●	○	○
Department of Justice	●	○	○
Department of Labor	●	○	○
Department of the Navy	●	○	○
Department of State	●	○	○
Department of Transportation	●	◑	○
Department of the Treasury	●	◑	◑
Department of Veterans Affairs	◑	○	○
Environmental Protection Agency	●	○	○
General Services Administration	●	◑	○
National Aeronautics and Space Administration	●	○	○
National Science Foundation	●	○	○
Nuclear Regulatory Commission	●	◑	○
Office of Personnel Management	●	◑	○
Small Business Administration	●	○	○
Social Security Administration	●	○	○
US Agency for International Development	●	●	●

Source: GAO analysis of agency data.

● Satisfied ◑ Partially Satisfied ○ Not Satisfied

Almost All Agencies Had Defined the Goals or Purposes of Their Enterprise Architecture

Before an agency knows what outcomes it should measure, it needs to define the purpose or expected value (i.e., goals) of its architecture. The purpose can include, among other things, consolidating the organization's IT infrastructure, normalizing and integrating its data and promoting information sharing, reengineering core business or mission functions and processes, modernizing applications and sharing services, modernizing the entire IT environment, and transforming how the organization operates. Expected value from implementation of enterprise architecture can include, for example, reduced operating costs, enhanced ability to quickly and less expensively change to meet shifting external environment and new business demands or opportunities, or improved alignment between operations and strategic goals. Twenty-six of the agencies we reviewed had fully defined their architecture goals or purposes. The following are examples of the goals or purposes defined by these agencies:

- The Department of Energy's goals include identifying, reusing, and leveraging, where possible, existing and planned technology and infrastructure components across the department and identifying areas, through capital planning and investment control and enterprise architecture integration analysis, to reduce costs, identify redundancy, and increase system and process effectiveness.

- HHS's goals include enabling improved mission and business outcomes by providing products to support sound decisions, business processes, and effective solutions; enabling the optimized use of resources; and increasing interoperability and information sharing within HHS and between HHS and external stakeholders.

- HUD's goals include improving the efficiency and effectiveness of the department's programs; simplifying its IT environments by promoting standards and sharing and reusing common technologies; improving interoperability by establishing enterprise-wide standards; and reducing system development and operation and maintenance costs by eliminating duplicative investments, promoting sharing of common services, and establishing department-wide standards.

- The Department of Justice's purpose includes identifying redundant legacy programs to either retire or migrate to an enterprise solution, thereby reducing the complexity and cost of the IT environment.

- The Department of Labor's purpose is to use its enterprise architecture process with its capital planning and investment management process to ensure that investments support strategic goals and are not duplicative of existing business solutions. Using this approach, the department plans to identify duplicative resources and investments, gaps, and opportunities for internal and external collaboration resulting in operational improvements and cost-effective solutions to business requirements.

- The Department of Transportation's goal is to use its architecture as a decision-making tool to support business plan development and identify areas of duplication and inefficiencies in the department.

- The General Services Administration's (GSA) goals are to increase system interoperability and cost efficiencies, reduce duplication, and increase innovation.

- The National Science Foundation's goals include improving utilization of IT resources by eliminating duplicative investments and promoting the sharing of common services and standards.

- The U.S. Agency for International Development's (USAID) goals include facilitating analysis of the agency's IT environment, including IT hardware, software, and enterprise applications, to promote the effective and efficient deployment of IT services.

However, one agency (Department of Veterans Affairs) had only partially defined its architecture's purpose. The Department of Veterans Affairs (VA) is in the process of developing an architecture program overview statement and guiding principles. Specifically, according to draft documentation, the department's architecture is to guide efficient, effective, and interoperable implementation of the department's vision of providing seamless delivery of benefits and services to veterans. According to department officials, these architecture principles are expected to be finalized and formally released by September 30, 2012.

Because they had defined the goals or purposes of their architectures, almost all of the agencies had taken an important first step toward establishing metrics and a method for measuring architecture outcomes and benefits.

More than Half of the Agencies Had Not Yet Established Metrics and a Method for Measuring Enterprise Architecture Value

Measuring the extent to which the expected value is actually being realized is important to identifying what, if any, architecture program changes are warranted. According to our framework, agencies should establish measurable, meaningful, repeatable, consistent, and actionable metrics that align with the architecture's intended purpose or strategic goals and document a methodology that provides the steps to be followed to consistently and repeatedly measure architecture outcomes and benefits. Further, according to OMB guidance, metrics should measure outcomes (i.e., results of products and services such as benefits to Congress and the American taxpayer), or expected value, rather than output (i.e., direct products and services).

Of the federal agencies that we reviewed, three had fully established metrics and a method to measure architecture outcomes and benefits, while eight had partially done so. Specifically, HHS, HUD, and USAID had fully established metrics and a method for measuring and reporting enterprise outcomes and benefits.

- HHS had established a metric to measure the extent to which it increases the number of services that are reused based on its enterprise architecture service component reference model.[30] The department had also established a method for how the metrics are to be measured, including how they are to be calculated, the data sources to be used, and targets to be achieved.

- HUD had established a method and metrics to measure the extent to which the department decreases the number of technology products that duplicate existing capabilities and the extent to which it has decreased the number of obsolete systems in its IT inventory, using its enterprise architecture. The department had also established the steps to measure results and outcomes, including identifying appropriate sources, and determining baseline, target, and actual value measurements.

---

[30]A service component reference model identifies and classifies IT service components.

- USAID had established metrics and guidance for measuring enterprise architecture outcomes, including cost savings and avoidance due to process efficiency, technology standardization, retirement, and consolidation.

Partial steps had been taken by the other eight agencies. Specifically, one agency (Army) had established a metric and method—for measuring the extent to which it reduces the number of applications within data centers— but only for one of its three segment architectures. The other seven agencies had established metrics but not a method for measuring and reporting architecture outcomes and benefits. The metrics for each of the seven agencies are described below.

- Commerce established as a metric the IT cost reduction associated with adopting enterprise-wide standards.

- Education established a metric to measure spending on development, modernization, and enhancement relative to steady-state spending (i.e., the cost to maintain current systems and technologies).

- Transportation established an expected architecture outcome of reduced total cost of ownership of IT investments, and planned to measure cost savings and/or cost avoidance identified through reviews of business processes, data, applications, and technology.

- Treasury established architecture metrics associated with its data center consolidation initiative, including the extent to which it decreases the number of servers, increases the percentage of operating systems that are virtual, and decreases the demand for data center square footage.

- GSA established as a metric the extent to which the agency is increasing its use of IT standards.

- NRC established a metric to measure progress toward having common access controls by measuring the reduction in passwords and sign-ons.

- Office of Personnel Management (OPM) has established cost savings as a metric to measure architecture outcomes.

The remaining 16 agencies in our study had not established metrics or a method for measuring architecture outcomes. While some of these agencies had established metrics that measure output, such as the percentage or number of segments and solution architectures or architecture artifacts that have been reviewed and approved by the enterprise architecture program, these metrics do not measure outcomes (i.e., results of enterprise architecture products and services such as benefits to Congress and the American taxpayer) of the program.

Without established metrics and a method to measure architecture outcomes, agencies cannot ensure that they are able to consistently and repeatedly measure outcomes.

Five Agencies Had Fully or Partially Measured and Reported Architecture Value

Using established metrics and a documented method, architecture outcomes should be periodically measured and reported to senior executives. We have previously found that executive leadership was key to addressing management challenges identified by enterprise architecture programs, such as overcoming limited executive understanding and inadequate funding. As such, architecture outcomes and benefits should be periodically reported to senior agency executives who are responsible for making decisions about the architecture program and whether to invest additional resources or make changes to the program.

Of the 27 agencies in our review, 1 had consistently and repeatedly measured and reported, using established metrics, outcomes of its architecture program. Specifically, USAID had reported monthly the measured outcomes to its CIO and through an internal agency website established for CIO staff. Outcomes reported include cost savings of $12.3 million and cost avoidance of $9.5 million as a result of transitioning disparate human resource systems to a human resource shared services center using enterprise architecture. The agency also reported estimated savings of $15.7 million from moving its e-mail service to a cloud-based solution, which was recommended by the architecture team to replace multiple installations of the current e-mail solution.

Two other agencies had measured and reported outcomes with an established method, but did so only once. Specifically,

- HHS determined, based on its enterprise architecture service component reference model, and reported to the CIO in November 2010 that 16 percent of its services were reused. However, the department had not measured the metric again and thus did not know the extent to which it had increased its reuse of services since 2010.

- HUD measured and submitted its architecture value measurement report for fiscal year 2011 to a department executive committee in August 2012, and highlighted areas, based on measurements, where additional focus and improvement are needed. For example, the report noted that HUD had not decreased the number of technology products in its enterprise architecture technical reference model[31] that duplicate existing capabilities.

While two additional agencies, as described below, had also measured and reported architecture outcomes once, they did so without an established method for measuring outcomes, but rather in an ad hoc manner.

- Education reported in its October 2011 Office of the CIO Organization Performance Review report that development, modernization, and enhancement funding in the IT portfolio increased from 10 percent of total IT spending in fiscal year 2011, to 13 percent of total IT spending in fiscal year 2012 through use of the department's architecture segment modernization planning process. However, Education had not established a method for measuring and reporting architecture outcomes and benefits. As a result, it cannot ensure that it will be able to consistently and repeatedly measure architecture outcomes over time.

- Treasury reported in its E-Government Act Report for fiscal year 2011 that its enterprise architecture plans focused on reducing duplication through its data center consolidation initiative. Accordingly, it reported through its CIO to OMB a reduction of 1,283 in the number of servers, an increase from 25 percent to 36 percent of operating systems that were virtualized, and a reduction in data center square footage of 15,896 between 2010 and 2011. However, Treasury had not established a method for measuring and reporting architecture outcomes and benefits. As a result, it cannot ensure that it will be able to consistently and repeatedly measure outcomes over time.

The remaining agencies (22) had not yet measured and reported architecture outcomes to senior executives. Agencies generally cited two reasons why they had not done so. Specifically, agencies had not

---

[31]A Technical Reference Model describes the standards, specifications, and technologies that support the delivery of service components.

determined how to attribute discrete outcomes to enterprise architecture when other activities, such as strategic planning, capital planning, and project management may have contributed to the outcomes. In addition, agencies cited an absence of guidance and best practices for how to measure enterprise architecture outcomes. As discussed in the next section, OMB has issued recent enterprise architecture guidance to agencies, but has not yet provided sufficient details on the method and metrics that could be used to measure architecture program outcomes.

Collectively, this means that while efforts are underway, without the use of associated measures and metrics by the majority of agencies, the 27 major departments and agencies are not positioned to know whether outcomes and benefits are in fact being achieved. Until agencies establish an approach for measuring enterprise architecture outcomes, including a documented method (i.e., steps to be followed) and metrics that are measurable, meaningful, repeatable, consistent, actionable, and aligned with the agency's enterprise architecture's strategic goals and intended purpose; and measure and report enterprise architecture outcomes and benefits to top agency officials and to OMB, agency senior executives are less likely to be sufficiently informed about whether to invest additional resources or make changes to the enterprise architecture program.

OMB's Guidance to Agencies Lacks Sufficient Details on Measuring Enterprise Architecture Value

The E-Government Act of 2002 assigned OMB the responsibilities for overseeing the development of enterprise architectures within and across the federal agencies. Since then, OMB has issued guidance and frameworks for developing and using architectures, including a May 2012 policy and guidance on establishing a common approach to developing and using enterprise architectures within and between federal agencies. The policy required each federal agency to submit by August 31, 2012, an enterprise roadmap that reports, among other things, how architecture program effectiveness and efficiency will be measured. However, while this guidance begins to describe an approach for collaboratively identifying, planning for, achieving, and measuring needed organizational outcomes (called the Collaborative Planning Methodology) and discusses the difference between outcome and output measures, it does not provide sufficient details on the method and metrics that could be used to measure architecture program outcomes.

As we noted earlier, according to our framework a methodology should provide the steps to be followed to consistently and repeatedly measure architecture outcomes and benefits. While OMB's collaborative planning methodology emphasizes the importance of measuring benefits and describing how they will be measured when planning for and executing collaborative projects, it does not call for specific metrics to be used or identify steps to be followed to consistently and repeatedly measure outcomes and benefits. Further, OMB does not call for agency roadmaps to include measurement methods and metrics, and reports on specific outcomes and benefits that an agency has achieved or plans to achieve.

In discussing this matter, the Federal Chief Enterprise Architect agreed with our assessment but stated that the methodology was not intended to be guidance on measuring architecture value and that more detailed guidance was being developed. According to the Federal Chief Architect, the detailed guidance on measuring enterprise architecture value is expected to be provided to agencies by December 2012, in time to facilitate the development of their next roadmap submissions, due in April 2013.

With the development of clear and sufficiently detailed guidance on measuring outcomes by OMB, agencies may be better positioned to develop methods and metrics for measuring and reporting the strategic value produced by their enterprise architecture programs. Moreover, with reports about architecture outcomes and benefits, agency executives could increase their understanding of the architecture programs, such that warranted changes could be addressed, or the need for expanded architecture development and use may be able to be economically justified. An established method and metrics to measure outcomes and benefits will enable agencies to repeatedly and consistently measure and report the extent to which they are achieving value.

Conclusions

Enterprise architecture value has yet to be measured and reported across the majority of the federal agencies. While most of the agencies reviewed have defined their architecture's goals or purpose, the majority had yet to establish metrics and a method for measuring and reporting architecture value. This means that while efforts are underway, the majority of the agencies do not know the extent to which they are realizing benefits that they have set out to achieve, such as cost savings or avoidance through eliminating duplicative investments. Furthermore, most of the agencies had not measured and reported outcomes to stakeholders or agency executives. Without measurable, meaningful, repeatable, consistent, and actionable metrics that align with the architecture's strategic goals or intended purpose and a documented methodology that provides the steps to be followed to consistently and repeatedly measure outcomes and benefits, senior agency executives may not have the information needed to determine whether to invest additional resources or make changes to the program. OMB's forthcoming guidance is an opportunity to overcome the absence of detailed directions to agencies on how they can measure and report enterprise architecture strategic value.

Recommendations for Executive Action

To enhance federal agencies' ability to realize enterprise architecture benefits, we recommend the following actions.

We recommend that the Secretaries of the Departments of Agriculture, the Air Force, the Army, Commerce, Defense, Education, Energy, Homeland Security, the Interior, Labor, the Navy, State, Transportation, the Treasury, and Veterans Affairs; the Attorney General; the Administrators of the Environmental Protection Agency, General Services Administration, National Aeronautics and Space Administration, and Small Business Administration; the Commissioners of the Nuclear Regulatory Commission and Social Security Administration; and the Directors of the National Science Foundation and the Office of Personnel Management ensure the following two actions are taken:

- fully establish an approach for measuring enterprise architecture outcomes, including a documented method (i.e., steps to be followed) and metrics that are measurable, meaningful, repeatable, consistent, actionable, and aligned with the agency's enterprise architecture's strategic goals and intended purpose; and

- periodically measure and report enterprise architecture outcomes and benefits to top agency officials (i.e., executives with authority to commit resources or make changes to the program) and to OMB.

In addition, we recommend that the Secretaries of the Departments of Health and Human Services and Housing and Urban Development ensure that enterprise architecture outcomes are periodically measured and reported to top agency officials.

To assist agencies in measuring and reporting outcomes achieved through enterprise architecture, we recommend that the Director of OMB ensure that the planned December 2012 guidance for enterprise architecture value measurement and reporting includes

- sufficient details on the method and metrics that agencies could use to measure their architecture program's value and

- a requirement for agencies to include in their April 2013 enterprise roadmap submissions a measurement method (i.e., steps to be followed) and metrics, and report on the outcomes and benefits achieved through enterprise architecture.

Agency Comments and Our Evaluation

We received comments on a draft of this report from OMB and the 24[32] agencies in our study. OMB's Federal Chief Enterprise Architect stated in oral comments and via e-mail that OMB agreed with the report and the recommendations. Among the agencies in our study, 5 responded via e-mail that they had no comments on our draft report.[33] One of these agencies—USAID—provided technical comments, which we incorporated as appropriate. An additional 2 agencies provided letters stating that they had no comments on our draft report. Specifically, Labor's Assistant Secretary for Administration and Management stated in a written response (reproduced in appendix III) that the department had no comments, and Treasury's Deputy Assistant Secretary for Information Systems and Chief Information Officer stated in a written response (reproduced in appendix IV) that the department had no comments on the draft report but appreciated GAO's efforts in its development.

Among the remaining agencies, 13 agreed with our results. These comments are summarized below.

- USDA's Acting Chief Information Officer stated in written comments that the department concurred with our findings and recommendations and plans to develop metrics and guidance to comply with OMB guidance on measuring enterprise architecture, when it is provided. USDA's written comments are reproduced in appendix V.

- Commerce's Acting Secretary stated in written comments that the department agreed with the general findings and specific recommendations as they relate to the department. Commerce's written comments are reproduced in appendix VI.

- DOD's Deputy CIO for Information stated in written comments that the department concurred with our recommendations and is developing

---

[32] DOD included comments from the departments of the Air Force, Army, and Navy.

[33]Transportation, GSA, NSF, and NRC and USAID.

an enterprise architecture management plan that provides high-level processes, including to measure architecture outcomes. DOD's written comments are reproduced in appendix VII.

- Education's CIO stated in written comments that the department concurred with our recommendations and described steps the department plans to take to address the recommendations. For example, the department plans to develop, document, and implement a measurement and reporting method that will be used to periodically monitor its progress toward achieving goals, desired outcomes, and benefits. The department also provided technical comments that we have incorporated, as appropriate, in the report. Education's written comments are reproduced in appendix VIII.

- DHS's Director, Departmental GAO-OIG Liaison Office, stated in written comments that the department concurred with our recommendations and described actions it plans to take to address them. For example, DHS stated that it plans to brief architecture outcomes for the goals and objectives outlined in the strategic plan to the CIO by October 31, 2012. DHS's written comments are reproduced in appendix IX.

- Interior's Assistant Secretary for Policy Management and Budget stated in written comments that the department concurred with our recommendations. Interior's written comments are reproduced in appendix X.

- A Management Analyst in Justice's audit liaison group commented via e-mail that the department agreed with our recommendations. The official also provided technical comments that we have incorporated, as appropriate.

- State's Comptroller provided written comments which noted that the department concurred with our conclusions and recommendations, and described steps being taken or planned to address the recommendations. For example, the department plans to implement, in fiscal year 2013, a metric to measure reduction in the percentage of information exchange elements between critical management systems through use of its enterprise architecture. State's written comments are reproduced in appendix XI.

- VA's Chief of Staff stated in written comments that the department generally agreed with our conclusions and concurred with the recommendations. The department also described actions it had taken in fiscal year 2012 to re-establish its enterprise architecture program and actions it plans to take to continue to mature the program in fiscal year 2013 that would begin to address our recommendations. VA's written comments are reproduced in appendix XII.

- NASA's CIO stated in written comments that the agency concurred with the recommendations, and described steps the agency plans to take to address them. For example, NASA plans to revise, by June 2013, its procedural requirements to better align architecture metrics and methods to measure outcomes. NASA's written comments are reproduced in appendix XIII.

- SSA's Deputy Chief of Staff stated in written comments that the agency agreed with the recommendations. SSA's written comments are reproduced in appendix XIV.

- EPA's Assistant Administrator and Chief Information Officer stated in written comments that the agency agrees with our findings and described steps it plans to take to address our recommendations. For example, it plans to develop a performance measurement plan, which will identify processes to measure enterprise architecture outcomes. EPA's written comments are reproduced in appendix XV.

- HHS's Assistant Secretary for Legislation in written comments stated that the department concurred with our findings and described actions it is taking, and plans to take, to improve architecture value measurement. HHS's written comments are reproduced in appendix XVI.

The remaining four agencies provided comments that expressed concerns with certain aspects of our results. These comments are summarized below.

- Energy's Chief Architect provided written comments in which the department stated that it had established metrics and a method for measuring architecture value and that its efforts justify a partially-satisfied rating. However, our study found that, although the department has submitted its Enterprise Modernization Roadmap to OMB, the roadmap includes potential architecture program metrics that are still being defined and have yet to be finalized and approved. The department also stated that it had achieved several accomplishments which justified a partially-satisfied rating for measuring and reporting architecture outcomes and benefits. In this regard, the department highlighted accomplishments such as collecting and reporting architecture success stories to a working-sub group of its Information Technology Council, which includes senior-level IT management from the offices of the Chief Financial Officer and the CIO. However, we found that these accomplishments and success stories are not based on an established set of metrics and a documented, consistently applied methodology for measuring and reporting architecture outcomes. As a result, the department cannot ensure that it will be able to consistently and repeatedly measure outcomes over time. Thus, we stand by our findings. Energy's written comments are reproduced in appendix XVII.

- HUD's CIO provided written comments on the report stating that the department has complied with our recommendation that enterprise architecture outcomes be periodically measured and reported to top agency officials. Specifically, the department stated the fiscal year 2011 report on outcomes was submitted to an executive committee in August 2012. It also stated that an EA Value Measurement Plan will be issued annually and results of the measures in the plan will be documented in an annual report for the fiscal year. However, while the department has completed and submitted its first report (i.e., for fiscal year 2011) to an executive committee, it has yet to measure and report on the metrics again, and therefore does not know the extent to which it has achieved its target outcomes. In addition, the official commented on the statement in the background of our report that the department had not yet finalized its architecture policy, as we had previously recommended. The official commented that a policy has been in place since April 2002. However, as the official stated, its updated policy has yet to be approved. As a result, we stand by the statement. HUD's written comments are reproduced in appendix XVIII.

- A Senior Analyst, e-mailing on behalf of OPM's Office of the CIO, provided comments in which the agency stated that savings through enterprise architecture are being measured and reported. However, it provided no evidence to support this statement and stated that more information will be available once the department implements a revised enterprise architecture roadmap, expected by the end of December 2012. As a result, we did not change our finding.

- A Program Manager in SBA's Office of Congressional and Legislative Affairs provided e-mail comments. Specifically, in comments on our finding that stated the agency had not defined its enterprise architecture strategic goals or intended purpose, SBA stated that its architecture's purpose and goals are defined and it provided supporting documentation in this regard. In response, we updated the finding and recommendation accordingly. In comments on our finding that the agency had not established a method or metrics to measure outcomes and benefits, SBA provided its Capital Planning and Investment Control Policy Guide and its fiscal year 2011 Summary of Performance and Financial Information. However, neither of these documents demonstrated a method and metrics for measuring architecture outcomes. In comments on our finding that the agency is not periodically measuring architecture outcomes and benefits, the agency stated that outcomes are measured and reported as part of the integrated enterprise architecture-capital planning and investment control effort through the Business Technology Investment Advisory Committee and the Business Technology Investment Council. While the agency provided some documentation, it did not provide requested examples of reports submitted to the officials. The agency also added that outcomes are reported in its annual performance report and provided documentation. While we agree that some outcomes are documented, the report does not highlight architecture-related outcomes, and SBA did not provide documentation linking the outcomes to its enterprise architecture. Therefore, we did not change our findings relative to establishing a method and metrics and measuring and reporting architecture outcomes.

We are sending copies of this report to other interested congressional committees; the Director of the Office of Management and Budget; the Secretaries of Agriculture, the Air Force, the Army, Commerce, Defense, Education, Energy, Health and Human Services, Homeland Security, Housing and Urban Development, the Interior, Labor, the Navy, State, Transportation, the Treasury, and Veterans Affairs; the Attorney General; the Administrators of the Environmental Protection Agency, General Services Administration, National Aeronautics and Space Administration, Small Business Administration, and U.S. Agency for International Development; the Commissioners of the Nuclear Regulatory Commission and the Social Security Administration; and the Directors of the National Science Foundation and Office of Personnel Management. In addition, the report will be available at no charge on the GAO website at http://www.gao.gov.

If you or your staffs have questions on matters discussed in this report, please contact me at (202) 512-6304 or melvinv@gao.gov. Contact points for our Offices of Congressional Relations and Public Affairs may be found on the last page of this report. GAO staff who made contributions to this report are listed in appendix XIX.

Valerie C. Melvin
Director
Information Management and
   Technology Resources Issues

Appendix I: Objective, Scope, and Methodology

Our objective was to determine the extent to which federal agencies are measuring and reporting enterprise architecture outcomes and benefits. To accomplish the objective, we focused on 28 enterprise architecture programs relating to 27 major departments and agencies. These 27 included the 24 departments and agencies identified in the Chief Financial Officers Act,[1] as well as the Departments of the Air Force, Army, and Navy. At the Department of Defense (DOD), we reviewed two department-wide architecture programs—the Business Enterprise Architecture and the DOD Enterprise Architecture. Table 2 identifies the agencies included in our study. These agencies were also included in our 2006 review of agencies' management maturity.[2]

Table 2: Agencies Included in Our Study

Agency
Department of Agriculture
Department of the Air Force
Department of the Army
Department of Commerce
Department of Defense (Business Enterprise Architecture and Enterprise Architecture)
Department of Education
Department of Energy
Department of Health and Human Services
Department of Homeland Security
Department of Housing and Urban Development
Department of the Interior
Department of Justice
Department of Labor
Department of the Navy
Department of State
Department of Transportation
Department of the Treasury
Department of Veterans Affairs

[1] See 31 U.S.C. § 901(b).

[2] GAO, *Enterprise Architecture: Leadership Remains Key to Establishing and Leveraging Architectures for Organizational Transformation*, GAO-06-831 (Washington, D.C.: Aug. 14, 2006).

Agency
Environmental Protection Agency
General Services Administration
National Aeronautics and Space Administration
National Science Foundation
Nuclear Regulatory Commission
Office of Personnel Management
Small Business Administration
Social Security Administration
United States Agency for International Development

Source: GAO.

We reviewed the responses to a survey we administered in May 2011, of federal agencies' efforts to measure and report enterprise architecture results and outcomes. The purpose of the survey was to follow up with the agencies we reviewed in 2006, about the costs and benefits associated with their enterprise architecture programs.[3] In addition, we requested and reviewed documents describing each agency's enterprise architecture program, focusing on the purpose and goals of the programs and the methods and metrics used to measure outcomes, such as IT strategic plans, program management plans, enterprise transition plans, enterprise modernization roadmaps, and value measurement plans. We analyzed the extent to which the documentation satisfied elements related to outcomes measurement and reporting in version 2.0 of our Enterprise Architecture Management Maturity Framework (EAMMF).[4] Specifically, we assessed agencies against elements of the framework related to defining the architecture's intended purpose or strategic goals, establishing a method and metrics to measure architecture strategic mission value (outcomes and benefits), and periodically measuring and reporting outcomes and benefits to an architecture executive committee. We also reviewed outcomes and benefits reported to agency architecture oversight officials, for example, in value measurement reports or performance measurement reports. We assessed the reliability of the reported outcomes and benefits by discussing with agency officials the

---

[3]GAO-06-831.

[4]GAO, *Organizational Transformation: A Framework for Assessing and Improving Enterprise Architecture Management (Version 2.0)*, GAO-10-846G (Washington, D.C.: August 2010).

method and data used to determine them, and by reviewing relevant documents, such as business cases and return on investment analyses.

To guide our analysis, we defined detailed evaluation criteria for determining whether a given element was fully satisfied, partially satisfied, or not satisfied. To fully satisfy an element, sufficient documentation had to be provided to permit us to verify that all aspects of the element were met. To partially satisfy an element, sufficient documentation had to be provided to permit us to verify that at least some aspects of the element were met. Elements that were neither fully nor partially satisfied were judged to be not satisfied.

Our evaluation included independently analyzing the extent to which each agency had satisfied the elements using the survey responses and supporting documentation as a starting point. We then corroborated the analyses with supporting documentation, sought additional information as necessary through interviews with the agencies' architecture officials, obtained and reviewed additional documentation as appropriate, and refined our determinations about the degree to which each element was satisfied. Finally, we shared with agencies preliminary versions of the analyses that appear in this report as appendix II, and made further adjustments, as appropriate, based on additional discussions and supporting documentation. We also met with the Federal Chief Enterprise Architect to discuss current efforts and plans to guide federal agencies' efforts to measure and report enterprise architecture outcomes and benefits.

We conducted our work from November 2011 to September 2012 in accordance with generally accepted government auditing standards. Those standards require that we plan and perform the audit to obtain sufficient, appropriate evidence to provide a reasonable basis for our findings and conclusions based on our audit objective. We believe that the evidence obtained provides a reasonable basis for our findings and conclusions based on our audit objective.

Appendix II: Detailed Assessments of Individual Departments and Agencies against Relevant Elements of Our Enterprise Architecture Management Maturity Framework

The following sections summarize the extent to which each of the 27 departments and agencies addressed elements in GAO's Enterprise Architecture Management Maturity Framework (EAMMF) that pertain to measuring and reporting enterprise architecture outcomes and benefits.

The assessments given for each element are defined as follows:

● The agency or department fully satisfied the element.

◖ The agency or department satisfied some, but not all, aspects of the element.

○ The agency or department did not satisfy any aspect of the element.

Department of Agriculture

Table 3 shows the Department of Agriculture's (USDA) satisfaction of relevant framework elements in version 2.0 of GAO's EAMMF.

Table 3: Department of Agriculture Satisfaction of EAMMF Elements

Element	Satisfied?	Summary
The enterprise architecture's intended purpose or strategic goals are defined.	●	USDA has defined its architecture's purpose and goals. Specifically, according to the department's IT strategic plan for 2012 through 2016, enterprise architecture and portfolio management practices are to be used to address mission needs in a cost-effective and efficient manner. In addition, according to the plan, the enterprise architecture program is to be used as a strategic enabler to drive planning activities, provide insights, and identify improvement opportunities for consolidation and reuse.
A method and metrics have been established to measure enterprise architecture strategic mission value (outcomes and benefits).	○	The department has not established a method and metrics for measuring enterprise architecture outcomes; however, according to agency officials, it plans to do so. In particular, officials said the department is planning to integrate the capital planning, budget, and enterprise architecture processes and is working with the Federal Chief Information Officer (CIO) in the area of shared services using architecture. In addition, according to the department's IT strategic plan, the department had planned to establish and periodically report on enterprise architecture program metrics by the end of fiscal year 2012. Officials explained that they now expect to complete this effort by the end of fiscal year 2013 because they are waiting for the Office of Management and Budget to provide additional guidance.
Enterprise architecture outcomes and benefits are periodically measured and reported to the agency's enterprise architecture executive committee.	○	The department has not periodically measured and reported enterprise architecture outcomes.

Source: GAO analysis of agency-provided data.

Department of the Air Force

Table 4 shows the Department of the Air Force's satisfaction of relevant framework elements in version 2.0 of GAO's EAMMF.

Table 4: Department of the Air Force Satisfaction of EAMMF Elements

Element	Satisfied?	Summary
The enterprise architecture's intended purpose or strategic goals are defined.	●	The December 2009 Air Force Architecting Concept of Operations defines the vision and goal for the department's architecture as follows:
		• Vision: to enable the delivery of timely, relevant, unambiguous information to support informed decision making by Air Force leaders to maximize military capabilities while optimizing allocation of resources.
		• Goal: to use architecture to unravel the complexity of systems, processes, and programs to reveal their interdependent relationships to decision makers, in an easily understandable format, so they may be adequately considered as decisions are made.
		Further, according to the Concept of Operations, the architecture is to be used as a tool to eliminate redundancy, build efficiency, and maximize resource distribution to ultimately increase the combat effectiveness of the Air Force.
A method and metrics have been established to measure enterprise architecture strategic mission value (outcomes and benefits).	○	Air Force officials reported that the department has not yet established a method or metrics to measure and report enterprise architecture outcomes and benefits. Officials stated that they have had a 60 percent architecting division personnel turnover rate since June 2011, and have not been able to identify industry-recognized enterprise architecture results metrics. Nonetheless, officials stated that they anticipate documenting potential metrics in October 2013.
Enterprise architecture outcomes and benefits are periodically measured and reported to the agency's enterprise architecture executive committee.	○	The Air Force has yet to measure and report enterprise architecture outcomes and benefits.

Source: GAO analysis of agency-provided data.

Department of the Army

Table 5 shows the Department of the Army's satisfaction of relevant framework elements in version 2.0 of GAO's EAMMF.

Table 5: Department of the Army Satisfaction of EAMMF Elements

Element	Satisfied?	Summary
The enterprise architecture's intended purpose or strategic goals are defined.	●	Army has defined the purpose and goals for each of its three segment architectures. According to Army officials, the collective purpose of the segment architectures is to make performance-based and cost-informed decisions that lead to the optimization of operations and technical environments. The purpose and goals for each of the three segment architectures has been defined as follows:

- **Generating Force.** According to Army's 2011 Business Transformation Plan, the purpose of this architecture is to drive integration across functional domains, ensure integration between the Generating and Operating Forces, and inform stakeholders on acquisition decisions pertaining to the migration of legacy functionality to Army's Enterprise Resource Planning solution. Expected benefits are the streamlining of end-to-end business processes aligned to the business enterprise architecture and the elimination or reduction of the need to tailor commercial-off-the-shelf systems.

- **Operating Force.** According to Army's 2004 Architecture Approval and Development memorandum, the purpose of this architecture is to assist in managing systems that support the current and future Army and to become a critical component in prioritizing and synchronizing Army-wide efforts. According to Army's 2011 Network Integration Roles, Responsibilities, and Functions memorandum, architecture analysis will be used to identify duplicative systems and incompatible implementations and to integrate requirements, platforms, and network capabilities across program offices, among other things.

- **Network.** According to Army's January 2011 Network Enterprise Architecture Foundation document, the purpose of this architecture is to provide relevant, trusted, affordable, and timely information to decision makers that support Army development and transformation, and help sustain the Army's transformation by facilitating an end-to-end alignment of capabilities and investments in support of Army planning and prioritization documents. Also, according to Army's 2010 Global Network Enterprise Construct Implementation Plan, the network architecture program's strategic initiatives include federating and integrating networks, enforcing standards, and aligning Army and federal data center consolidation initiative goals and objectives.

Element	Satisfied?	Summary
A method and metrics have been established to measure enterprise architecture strategic mission value (outcomes and benefits).	◑	Army has established metrics to measure outcomes and benefits of its Network segment architecture related to its data center consolidation initiative. The metrics measure the extent to which the number of data centers are closed each year, and the extent to which the number of servers, the amount of floor space, and energy usage and its associated costs are reduced. In addition, the Army has established a metric to measure the extent to which it reduces the number of applications on data servers and within data centers. It has also established, in its January 2012 Performance Plan for Reducing the Resources Required for Data Servers and Centers, a method for measuring the reduction in applications, which includes using an automated tool to collect, rationalize, and track the migration of its applications.
		However, Army has not established a method and metrics to measure outcomes and benefits for its Generating Force and Operating Force segment architectures. Although officials reported that they are tracking the status of architecture artifact development, artifacts are architecture program outputs rather than outcomes resulting from the use of an architecture.
		Army officials reported that the department faces a challenge that directly relates to the lack of a centralized enterprise architecture office that can provide oversight and guidance for architecture activities. A regulation intended to address this challenge, with measures for assessing whether the Army architecture is meeting the department's needs, has been drafted but has not been approved.
Enterprise architecture outcomes and benefits are periodically measured and reported to the agency's enterprise architecture executive committee.	○	Although the Network segment has established architecture method and metrics related to data center consolidation, it has yet to measure and report the outcomes and benefits. According to the Army's performance plan, an annual application reduction report will be provided to the DOD CIO starting in fiscal year 2013. With regard to the Generating Force and Operating Force segment architectures, the Army has yet to measure and report enterprise architecture outcomes and benefits.

Source: GAO analysis of agency-provided data.

| Department of Commerce | Table 6 shows the Department of Commerce's satisfaction of relevant framework elements in version 2.0 of GAO's EAMMF. |

Table 6: Department of Commerce Satisfaction of EAMMF Elements

Element	Satisfied?	Summary
The enterprise architecture's intended purpose or strategic goals are defined.	●	Commerce defined goals for its enterprise architecture program in its September 2010 Strategic Information Technology Plan for 2011-2015. These include using the department's enterprise architecture to continually improve its business processes, align resources with Commerce's top-level strategic goals, and identify and support key IT management decisions. According to the plan, Commerce plans to leverage its architecture to reduce redundancy in its IT portfolio, combine capabilities, utilize already-existing resources, and ensure that available IT resources are documented and visible for all potential users.
		According to the department's Chief Enterprise Architect, the department is reevaluating the goals and objectives of the enterprise architecture program to make it more responsive to management requirements and to place less emphasis on report and document generation. The department established an enterprise architecture objective to adopt enterprise-wide standards for enterprise architecture, purchasing, and cost savings in its balanced scorecard[a] for the first quarter of fiscal year 2012.
A method and metrics have been established to measure enterprise architecture strategic mission value (outcomes and benefits).	◑	Commerce has established metrics and associated targets to measure achievement of the objective to adopt enterprise-wide standards for enterprise architecture, purchasing, and cost savings. These include the number of IT product standards adopted (target is two) and IT cost reduction (target is $50,000 in annual savings).
		However, the department has yet to establish a methodology that provides the steps to be followed to measure enterprise architecture strategic mission value. According to the Chief Enterprise Architect, the department has documented a methodology to be used to demonstrate potential cost savings. However, officials did not provide supporting documentation.
		According to the Chief Enterprise Architect, measuring and reporting enterprise architecture outcomes is a challenge because it is difficult to attribute outcomes directly to architecture since outcomes are achieved through a larger process that includes strategic and capital planning.
Enterprise architecture outcomes and benefits are periodically measured and reported to the agency's enterprise architecture executive committee.	○	According to the Chief Enterprise Architect, the department achieved savings in the first quarter of 2012 by switching from a decentralized approach to procuring computers, software, and computer services to a single, department-wide vehicle. However, Commerce did not provide documentation to support the measurement and reporting of these cost savings.

Source: GAO analysis of agency-provided data.

[a]The balanced scorecard is a private-sector concept introduced by Robert Kaplan and David Norton in 1992 to assess organizational performance and is used by several government agencies. The balanced scorecard is a form of performance plan that is used to help measure performance, make improvements, and assess how well organizations are positioned to perform in the future.

Department of Defense– Business Enterprise Architecture

Table 7 shows the Department of Defense (DOD) Business Enterprise Architecture's (BEA) satisfaction of relevant framework elements in version 2.0 of GAO's EAMMF.

Table 7: DOD Business Enterprise Architecture Satisfaction of EAMMF Elements

Element	Satisfied?	Summary
The enterprise architecture's intended purpose or strategic goals are defined.	●	According to DOD's March 2012 BEA Overview and Summary Information, the purpose of the architecture is to (1) serve as a blueprint for business transformation that helps to ensure that the right capabilities, resources, and materiel are rapidly delivered to warfighters; (2) guide and constrain implementation of interoperable defense business system solutions; (3) guide information technology investments to align with strategic business capabilities; and (4) support portfolio management during the investment review process.
		DOD's September 2011 Strategic Management Plan for fiscal years 2012-2013, which establishes management goals for business operations, includes the goal to reengineer/use end-to-end business processes to reduce transaction times, drive down costs, and improve service. Associated with the goal is an initiative to improve business operations through optimal use of defense business systems and the BEA.
A method and metrics have been established to measure enterprise architecture strategic mission value (outcomes and benefits).	○	While the department has established metrics for measuring achievement of its goal to improve business operations through optimal use of defense business systems and the BEA, the metrics do not measure BEA outcomes and benefits. Specifically, DOD established metrics in its Strategic Management Plan for fiscal years 2012-2013 which include percentage of defense business systems/services represented in both the Defense Information Technology Portfolio Repository (the department's authoritative business systems inventory) and the BEA, percentage of defense business systems/services represented in both the Select and Native Programming Data Input System—Information Technology (the department's system used to prepare its budget submission) and the BEA, and percentage of defense business systems/services reporting to OMB through the BEA. However, these metrics measure output (i.e., direct products and services), rather than outcomes (i.e., results of enterprise architecture products and services such as benefits to Congress and the American taxpayer) of the enterprise architecture program.
		In addition, department officials stated that the department's process to measure and report architecture outcomes includes requiring components to submit examples of business system improvement for inclusion in the department's annual report to Congress on Defense Business Operations. These are to be substantiated with quantifiable measures that demonstrate desired business outcomes and benefits. However, the guidance provided to program offices for submitting these examples does not include the steps to be followed and metrics for measuring BEA outcomes.
Enterprise architecture outcomes and benefits are periodically measured and reported to the agency's enterprise architecture executive committee.	○	DOD has not periodically measured and reported enterprise architecture outcomes and benefits. Specifically, while the March 2012 Congressional Report on Defense Business Systems describes enhancements to the BEA related to business process modeling and standardizing business data, and reports the number of legacy systems that are not part of the target architecture (based on DOD IT Portfolio Repository data), the report does not include any additional examples.

Source: GAO analysis of agency-provided data.

Department of Defense– Enterprise Architecture

Table 8 shows DOD's Enterprise Architecture satisfaction of relevant framework elements in version 2.0 of GAO's EAMMF.

Table 8: Department of Defense Satisfaction of EAMMF Elements

Element	Satisfied?	Summary
The enterprise architecture's intended purpose or strategic goals are defined.	●	DOD has defined goals for its DOD enterprise architecture. Specifically, according to the department's February 2009 Directive on Management of the DOD Information Enterprise, the DOD enterprise architecture, which is composed of DOD enterprise and component levels, is to be maintained and applied to guide investment portfolio strategies and decisions, define capability and interoperability requirements, establish and enforce standards, guide security and information assurance requirements across DOD, and provide a sound basis for transition from the existing environment to the future.
A method and metrics have been established to measure enterprise architecture strategic mission value (outcomes and benefits).	○	DOD has yet to establish a method and metrics for measuring DOD enterprise architecture outcomes and benefits. According to officials, DOD's approach to establishing a method and metrics for measuring DOD enterprise architecture strategic mission value (outcomes and benefits) will be accomplished through the development and publication of a DOD instruction and an enterprise architecture management plan. In particular, the draft instruction on enterprise architecture calls for establishing metrics for assessing the effectiveness of the enterprise architecture to provide information that contributes to mission effectiveness and efficiency. In addition, the draft Enterprise Architecture Management Plan calls for the development of metrics to assess the use of enterprise architecture, provides examples of potential metrics, including reduction in redundancies in DOD's portfolio, and calls for the development of baseline and target threshold values for each selected metric. The plan also states that the DOD CIO and architecture organization are to determine the final set of metrics and threshold values based on the resources available to assess such metrics.
		However, the department has not yet issued the instruction or the plan or determined the specific method or final set of metrics to be used in measuring enterprise architecture outcomes and benefits. According to officials, the department expects the plan to be approved in December 2012 and the instruction to be approved in April 2013.
Enterprise architecture outcomes and benefits are periodically measured and reported to the agency's enterprise architecture executive committee.	○	DOD has yet to measure and report DOD enterprise architecture outcomes and benefits. According to DOD officials, the implementation of the instruction on enterprise architecture and the enterprise architecture management plan will allow the benefits of architecture to be measured and reported. However, the department has not yet issued the instruction or the plan.

Source: GAO analysis of agency-provided data.

Department of Education

Table 9 shows the Department of Education's satisfaction of relevant framework elements in version 2.0 of GAO's EAMMF.

Table 9: Department of Education Satisfaction of GAO EAMMF Elements

Element	Satisfied?	Summary
The enterprise architecture's intended purpose or strategic goals are defined.	●	Education has defined the purpose of its architecture program. Specifically, according to its July 2011 Enterprise Transition Plan, enterprise architecture provides for • a priority-driven approach to planning and executing the activities needed to transition from the baseline architecture to the target architecture, • improved strategic decision-making and communication to achieve the enterprise vision for technology at the department, • increased control mechanisms for technology planning and investment, • improved responsiveness to the enterprise technology needs of the department's business, and • the ability to leverage technology to create a more effective and efficient department.
A method and metrics have been established to measure enterprise architecture strategic mission value (outcomes and benefits).	◑	The department has established a metric to measure enterprise architecture outcomes, but has yet to establish a method. Specifically, according to the October 2011 Office of the CIO Organization Performance Review report, a key performance indicator for the Office of the CIO is to increase development, modernization, and enhancement (DME) spending through use of its enterprise architecture segment modernization planning process. To that end, the department has established a metric to measure the extent to which the ratio of the increase in spending on DME is increased relative to steady-state spending. According to the department's June 2010 IT Portfolio Analysis, increasing spending in DME leads to a decrease in spending to maintain current systems and technologies (i.e., steady-state spending). The department explained that its architecture program works with line-of-business segment owners to develop modernization plans that include achieving operational efficiencies. In June 2012, the department finalized an IT Portfolio Management Value Measurement methodology that describes the process for determining the value of an investment relative to the department's IT portfolio. The information is to be used to set priorities for funding decisions or selecting investments to be included in the department's IT portfolio. However, the process is not a method for measuring architecture strategic mission value.
Enterprise architecture outcomes and benefits are periodically measured and reported to the agency's enterprise architecture executive committee.	◑	In October 2011, the department reported the ratio of DME versus steady-state spending increased as a result of its enterprise architecture activities. Specifically, according to the department's October 2011 Office of the CIO Organization Performance Review report, DME funding in the IT Portfolio increased from 10 percent of total IT spending in fiscal year 2011, to 13 percent of total IT spending in fiscal year 2012 through use of the department's architecture segment modernization planning process. However, this metric has yet to be periodically measured and reported as an architecture outcome.

Source: GAO analysis of agency-provided data.

Department of Energy

Table 10 shows the Department of Energy's satisfaction of relevant framework elements in version 2.0 of GAO's EAMMF.

Table 10: Department of Energy Satisfaction of EAMMF Elements

Element	Satisfied?	Summary
The enterprise architecture's intended purpose or strategic goals are defined.	●	Energy has defined its enterprise architecture goals, which include: • Maintain alignment between technology solutions and department mission and goals. • Provide enterprise architecture training and outreach opportunities, thereby promoting enterprise architecture value and transparency to support better business decisions department-wide. • Work in conjunction with program and staff/support offices to further define, elaborate, and identify areas for additional development in line with the department's mission. • Continue to identify, reuse, and leverage, where possible, existing and planned technology and infrastructure components across the department. • Identify areas, through capital planning and investment control/enterprise architecture integration analysis, to reduce costs, identify redundancy, and increase system and process effectiveness. • Foster the organization and presentation of enterprise architecture to support decision making, program analysis, and efficient achievement of mission goals, utilizing an upgraded enterprise architecture data repository.
A method and metrics have been established to measure enterprise architecture strategic mission value (outcomes and benefits).	○	Energy has not established metrics and a method for measuring enterprise architecture strategic mission value. Specifically, although the department's August 2012 Enterprise Modernization Roadmap includes potential enterprise architecture program metrics (e.g., cost savings through retiring legacy systems and cost avoidance by leveraging existing solutions over procuring new ones through the use of enterprise architecture), the metrics are still being defined and have yet to be finalized and approved. Regarding a methodology, the roadmap states that appropriate processes will be developed once the metrics are developed and approved.
Enterprise architecture outcomes and benefits are periodically measured and reported to the agency's enterprise architecture executive committee.	○	The department has yet to measure and report enterprise architecture outcomes and benefits.

Source: GAO analysis of agency-provided data.

Department of Health and Human Services

Table 11 shows the Department of Health and Human Services' (HHS) satisfaction of relevant framework elements in version 2.0 of GAO's EAMMF.

Table 11: Department of Health and Human Services Satisfaction of EAMMF Elements

Element	Satisfied?	Summary
The enterprise architecture's intended purpose or strategic goals are defined.	●	HHS has established the following enterprise architecture goals and objectives: • Enable improved mission and business outcomes by providing products to support sound decisions, business processes, and effective solutions; providing structured methods and guidance; supporting the development of transformation plans for addressing business needs and priorities; enabling the optimized use of resources; and increasing interoperability and information sharing within HHS and between HHS and external stakeholders. • Provide a consolidated view of HHS's enterprise by providing a consolidated view of HHS's current and future business, information, and technologies; providing relevant, reliable, and timely information analytics capabilities to support sound business decisions; increasing the level of enterprise program integration and enterprise data sharing; facilitating the federated management and maintenance of enterprise architecture information through the use of a common framework; and facilitating the development of a consolidated view of information about systems and investments. • Strengthen the enterprise architecture program foundation by demonstrating the utility of enterprise architecture to support program and business needs and priorities, aligning the architecture program to HHS and federal enterprise business needs and priorities, and fostering effective enterprise architecture practices at the operating division level.
A method and metrics have been established to measure enterprise architecture strategic mission value (outcomes and benefits).	●	HHS has established a method and metrics to measure enterprise architecture outcomes. The department developed an Enterprise Architecture Value Measurement Plan in December 2009 which includes measuring the extent to which the department increases the number of service components that are reused. Specifically, it includes measuring the extent to which the department increases the percentage of applicable service components in its service component reference model that are provided by one IT system and used by another. The Enterprise Architecture Value Measurement Plan also identifies a method for how the metrics are to be measured, including how they are to be calculated, the data sources to be used, and targets to be achieved. In addition, the Enterprise Architecture Value Measurement Plan includes measuring potential cost avoidance based on recommendations made by the enterprise architecture program, such as for business process reengineering; elimination of redundant IT systems and services; and consolidation and reuse of IT systems, services, and data. However, agency officials said they have yet to develop a methodology for measuring cost avoidance. The Chief Enterprise Architect stated that it is a challenge to capture cost information, which is important to establishing a baseline, because investments cut across a number of systems.
Enterprise architecture outcomes and benefits are periodically measured and reported to the agency's enterprise architecture executive committee.	◑	Enterprise architecture results were measured and reported to the Chief Information Officer in November 2010. Among other measures, the briefing reported that 16 percent of service components were reused. However, the department did not measure the metric again and therefore, does not know the extent to which it increased its reuse of service components. According to the Chief Enterprise Architect, the department is in the process of establishing new IT and enterprise architecture priorities and intends to establish a new enterprise architecture results measurement and reporting approach by the end of fiscal year 2012.

Source: GAO analysis of agency-provided data.

Department of Homeland Security

Table 12 shows the Department of Homeland Security's (DHS) satisfaction of relevant framework elements in version 2.0 of GAO's EAMMF.

Table 12: Department of Homeland Security Satisfaction of EAMMF Elements

Element	Satisfied?	Summary
The enterprise architecture's intended purpose or strategic goals are defined.	●	The DHS Enterprise Architecture Strategic Plan for Fiscal Years 2012-2016 identifies the vision, mission, goals, and objectives of the department's enterprise architecture: • Vision: Improving mission and performance, optimizing resources, and unifying DHS. • Mission: To optimize DHS resources and provide a framework for strategic improvement and investment decisions. • Goals: Plan and execute segment architecture, enhance operating effectiveness, mature enterprise architecture practices, and strengthen enterprise architecture program management. • Objectives: To achieve each of the four goals, the plan outlines five corresponding objectives, including establishing performance metrics to drive standardization and accountability and identifying cost savings and avoidance through efficient and effective use of resources.
A method and metrics have been established to measure enterprise architecture strategic mission value (outcomes and benefits).	○	DHS has not established metrics and a method for measuring enterprise architecture outcomes and benefits. The department identified examples of enterprise architecture benefits (e.g., streamlined processes, increased tool reuse, cost avoidance, increased sharing, increased process improvements, and increased information sharing) and categorized them (direct user/customer benefits, operational/mission performance benefits, financial benefits, strategic/political benefits, and non-user/public benefits), and according to DHS officials, the examples and categories are being used to define enterprise architecture metrics. In addition, the department has developed a tool for documenting and reporting enterprise architecture outcomes. However, it has not finalized metrics and a method with detailed steps to ensure that outcomes are consistently and repeatedly measured.
Enterprise architecture outcomes and benefits are periodically measured and reported to the agency's enterprise architecture executive committee.	○	The department has yet to measure and report enterprise architecture outcomes and benefits. DHS officials stated that they expect to report architecture outcomes to a department executive body by October 1, 2012.

Source: GAO analysis of agency-provided data.

Department of Housing and Urban Development

Table 13 shows the Department of Housing and Urban Development's (HUD) satisfaction of relevant framework elements in version 2.0 of GAO's EAMMF.

Table 13: Department of Housing and Urban Development Satisfaction of EAMMF Elements

Element	Satisfied?	Summary
The enterprise architecture's intended purpose or strategic goals are defined.	●	According to agency documentation, the primary purpose of the enterprise architecture is to capture the information required to effectively plan a course for achieving HUD's strategic vision and goals. It is to be one element of interrelated planning activities that are to enable HUD managers and staff to define a vision, develop strategies and plans for achieving the vision, make resource decisions, implement strategies, and evaluate performance.
		HUD's enterprise architecture goals are to improve the efficiency and effectiveness of the department's programs; simplify HUD's IT environment by promoting standards and sharing and reusing common technologies; improve interoperability by establishing enterprise-wide standards; and reduce system development and operation and maintenance costs by eliminating duplicative investments, promoting sharing of common services, and establishing department-wide standards. The department also defined enterprise architecture goals in its draft 2011 Enterprise Architecture Value Measurement Plan, including the goal of enabling the use of enterprise IT technologies for reuse and to reduce infrastructure complexity. Associated with this goal are objectives to leverage existing IT technology products to meet business and functional requirements, standardize enterprise technologies where it is cost effective, and decommission obsolete systems that are no longer in use.
A method and metrics have been established to measure enterprise architecture strategic mission value (outcomes and benefits).	●	HUD has established a method and metrics to measure its enterprise architecture outcomes and benefits. Specifically, the department's fiscal year 2011 draft Enterprise Architecture Value Measurement Plan includes measuring the extent to which the department had decreased the number of technology products added to its Technical Reference Model (TRM) that duplicate existing capabilities, the extent to which it had increased the number of standardized enterprise technologies across the department that replace legacy products and do not duplicate existing capabilities, and the extent to which it had decreased the number of obsolete systems in its IT inventory. The plan also included steps to measure results and outcomes, including identifying appropriate sources, and determining baseline, target, and actual value measures.

Element	Satisfied?	Summary
Enterprise architecture outcomes and benefits are periodically measured and reported to the agency's enterprise architecture executive committee.	◑	The department measured and reported enterprise architecture outcomes for fiscal year 2011 to a department executive committee in August 2012. The December 2011 Enterprise Architecture Value Measurement Report highlights areas, based on measurements, where additional focus and improvement are needed. For example, according to the report, 12 technology products were added to its TRM that duplicate existing capabilities versus a target of 6; 0 enterprise-licensed technologies replaced legacy products versus a target of 3; and 0 obsolete systems were decommissioned versus a target of 13. According to officials, the department had not been able to retire these systems because their maintenance costs were included in fixed-price contracts that included systems that were currently being used. However, the department has yet to measure and report the metrics again, and therefore, does not know the extent to which it met its targets. According to the department, an architecture Value Measurement Plan will be issued annually and results of the measures in the plan will be documented in an annual report for the fiscal year.

Source: GAO analysis of agency-provided data.

Department of the Interior

Table 14 shows the Department of the Interior's (DOI) satisfaction of relevant framework elements in version 2.0 of GAO's EAMMF.

Table 14: Department of the Interior Satisfaction of EAMMF Elements

Element	Satisfied?	Summary
The enterprise architecture's intended purpose or strategic goals are defined.	●	The Department of the Interior established a purpose and goals for its enterprise architecture program in 2009. Specifically, the purpose is to develop, maintain, and oversee the implementation of an enterprise architecture that helps the department achieve its strategic goals. Goals for the enterprise architecture program are to (1) improve the implementation of architectural plans and (2) increase the portion of the enterprise architected. Department officials stated that an Enterprise Modernization Roadmap with an updated enterprise architecture purpose and goals is expected to be completed by September 2012.
A method and metrics have been established to measure enterprise architecture strategic mission value (outcomes and benefits).	○	While the department has established metrics for measuring achievement of its goals, the metrics do not measure enterprise architecture outcomes and benefits. According to the department's 2009 Enterprise Architecture Program Management Plan, the key performance measures for the program are the percentage of segments with completed architectures and the percentage of development/modernization/enhancement funding associated with completed and in-progress segment architectures. However, these are not measures of enterprise architecture program outcomes, but rather, measures of enterprise architecture development and implementation. Officials reported that an assessment of the department's enterprise architecture program was recently conducted and a new enterprise architecture program management plan is being developed; however, they have not established a time frame for when the plan will be completed.
Enterprise architecture outcomes and benefits are periodically measured and reported to the agency's enterprise architecture executive committee.	○	The department is not periodically measuring enterprise architecture outcomes and benefits.

Source: GAO analysis of agency-provided data.

Department of Justice

Table 15 shows the Department of Justice's satisfaction of relevant framework elements in version 2.0 of GAO's EAMMF.

Table 15: Department of Justice Satisfaction of EAMMF Elements

Element	Satisfied?	Summary
The enterprise architecture's intended purpose or strategic goals are defined.	●	Justice has defined its architecture's intended purpose. Specifically, according to its IT Strategic Plan for 2010 through 2015, the Enterprise Architecture Program Management Office is to review all IT investments to identify enterprise solutions that address the needs of a core mission area or multiple components. According to the strategy, enterprise solutions help to eliminate redundant IT investments, increase information sharing, and make use of shared infrastructure services, thus reducing the cost and complexity of managing the department's IT environment. Also, according to the strategy, enterprise architecture analysis is to support identifying redundant legacy programs to either retire or migrate to an enterprise solution, thereby further reducing the complexity and the cost of the IT environment.
A method and metrics have been established to measure enterprise architecture strategic mission value (outcomes and benefits).	○	The department has not established a method or metrics to measure enterprise architecture outcomes and benefits. According to the department's IT strategy, the enterprise architecture program is to help identify and eliminate redundant programs, thus reducing costs. However, the department stated that it is difficult to associate these cost savings specifically within the department-level enterprise architecture because a number of factors and groups contribute to the results.
Enterprise architecture outcomes and benefits are periodically measured and reported to the agency's enterprise architecture executive committee.	○	The department does not measure and report enterprise architecture outcomes and benefits.

Source: GAO analysis of agency-provided data.

Department of Labor

Table 16 shows the Department of Labor's satisfaction of relevant framework elements in version 2.0 of GAO's EAMMF.

Table 16: Department of Labor Satisfaction of EAMMF Elements

Element	Satisfied?	Summary
The enterprise architecture's intended purpose or strategic goals are defined.	●	Labor has defined the purpose of its enterprise architecture program. Specifically, according to its April 2011 Enterprise Transition Plan, the department uses its enterprise architecture process with its capital planning and investment management process to ensure that investments support strategic goals and are not duplicative of existing business solutions. Through use of this approach, according to the plan, the department is able to identify duplicative resources/investments, gaps, and opportunities for internal and external collaboration, resulting in operational improvements and cost-effective solutions to business requirements. In addition, the plan states that the department's enterprise architecture framework promotes interoperability and information sharing and provides benefits such as • enterprise target architecture definitions that support the department's mission objectives and strategic business plans, • identification of redundancy and consolidation opportunities, and • realization of cost savings and cost avoidance through improved performance.
A method and metrics have been established to measure enterprise architecture strategic mission value (outcomes and benefits).	○	The department has not established a method or metrics to measure its enterprise architecture outcomes and benefits. According to the April 2011 Enterprise Transition Plan, Labor will establish enterprise architecture program metrics to evaluate outcomes of the use of enterprise architecture in investment decision making. However, department officials reported that they have general measures related to capital planning that they use across the Office of the Chief Information Officer and they do not associate the measures specifically with enterprise architecture because a number of factors contribute to the results.
Enterprise architecture outcomes and benefits are periodically measured and reported to the agency's enterprise architecture executive committee.	○	The department does not measure and report enterprise architecture outcomes and benefits.

Source: GAO analysis of agency-provided data.

Department of the Navy

Table 17 shows the Department of the Navy's (DON) satisfaction of relevant framework elements in version 2.0 of GAO's EAMMF.

Table 17: Department of the Navy Satisfaction of EAMMF Elements

Element	Satisfied?	Summary
The enterprise architecture's intended purpose or strategic goals are defined.	●	DON has defined the purpose of its enterprise architecture program. According to the program's 2010 All-View document, which provides an overview and summary information of the DON enterprise architecture, the purpose is to: • Guide the department's investments towards achieving departmental goals and objectives. • Assist DON program managers in the development of their "solution architectures"–as mandated by the Joint Capabilities Integration Development System and Acquisition processes. More specifically, the enterprise architecture is to • promote interoperability; • delineate existing and future programs and projects; • establish uniform and standard models for business processes and IT systems that are common across DON; • document all aspects of the enterprise including the functional activities, business processes, information, participants, systems, applications, and supporting technology infrastructure; • support oversight and governance of IT investments; • enable and align business and IT investments through improved portfolio management, capital planning and investment control, and other acquisition and budgeting processes; • enable decision makers to identify capability gaps and overlaps; and • provide insight into Doctrine, Organization, Training, Materiel, Leadership, Personnel, and Facilities domains as they relate to the business, information, systems, applications, and information technology required for decision support at all levels of DON.
A method and metrics have been established to measure enterprise architecture strategic mission value (outcomes and benefits).	○	DON has not established a method or metrics to measure enterprise architecture outcomes and benefits. Officials stated that they anticipate establishing a method in the second half of 2013. Officials reported that a lack of best practices for measuring enterprise architecture value continues to inhibit their ability to demonstrate enterprise architecture return on investment.
Enterprise architecture outcomes and benefits are periodically measured and reported to the agency's enterprise architecture executive committee.	○	DON has yet to measure and report enterprise architecture outcomes and benefits.

Source: GAO analysis of agency-provided data.

Department of State

Table 18 shows the Department of State's satisfaction of relevant framework elements in version 2.0 of GAO's EAMMF.

Table 18: Department of State Satisfaction of EAMMF Elements

Element	Satisfied?	Summary
The enterprise architecture's intended purpose or strategic goals are defined.	●	According to State's IT Strategic and Tactical Plans for fiscal years 2011 to 2013, the purpose of the enterprise architecture is to focus on interoperability and application services.
A method and metrics have been established to measure enterprise architecture strategic mission value (outcomes and benefits).	○	The department has yet to establish a method or metrics for measuring enterprise architecture outcomes and benefits. Agency officials stated that while their objective is to have good IT investments, it is difficult to measure enterprise architecture's contribution because IT investment results are due to many factors, including good project management, and adequate funding, as well as enterprise architecture. While, according to the agency's IT Tactical Plan, a key performance indicator for its enterprise architecture program is evidence of increased use and value of enterprise architecture products and services in providing consistent and effective IT solutions, promoting interoperability, information sharing, and collaboration, State has yet to establish metrics and a method for measuring the value of its enterprise architecture products. Department officials reported that they expect to create metrics by December 2012.
Enterprise architecture outcomes and benefits are periodically measured and reported to the agency's enterprise architecture executive committee.	○	The department has yet to periodically measure and report enterprise architecture outcomes and benefits.

Source: GAO analysis of agency-provided data.

Department of Transportation

Table 19 shows Department of Transportation's satisfaction of relevant framework elements in version 2.0 of GAO's EAMMF.

Table 19: Department of Transportation Satisfaction of EAMMF Elements

Element	Satisfied?	Summary
The enterprise architecture's intended purpose or strategic goals are defined.	●	The department's Information Resources Management Strategic Plan for fiscal years 2007-2012 includes the goal to establish its enterprise architecture as the authoritative decision tool for IT investments. According to the plan, enterprise architecture is to be used as a decision-making tool to support business plan development, identify areas of duplication and inefficiencies in the department, and select top priorities for department-wide implementation and management.
A method and metrics have been established to measure enterprise architecture strategic mission value (outcomes and benefits).	◐	The department has established a metric but not a method to measure enterprise architecture strategic mission value. Specifically, according to Transportation's Information Resources Management Strategic Plan, an expected enterprise architecture outcome is reduced total cost of ownership, indicated by cost savings and/or cost avoidance identified through review of business processes, data, applications, and technology, and by the number/percentage of eliminated duplicative systems.
		In addition, in December 2008, the department developed an enterprise architecture Performance Measurement Guide, which included an enterprise architecture performance objective to support investment decisions for approved segment and solution architectures with an outcome of solutions that foster transparency, increase mission effectiveness, reduce redundancies, and minimize costs. However, a method to measure the objective was not established. Specifically, the guide included steps to finalize enterprise architecture performance measures and indicators, such as conducting outreach with stakeholders to disseminate information about performance measures and indicators, and capturing baselines and defining targets. In addition, the guide included a plan to collect and analyze data and to measure results on a quarterly basis, starting the first quarter of calendar year 2009. However, performance measures and indicators were never finalized and the guide did not include the steps to be followed to collect and analyze the data.
		According to the department's Chief Architect, enterprise architecture was a high priority in 2008 and 2009, but in mid-2010 the department shifted priorities and limited resources, which has constrained enterprise architecture efforts.
Enterprise architecture outcomes and benefits are periodically measured and reported to the agency's enterprise architecture executive committee.	○	The department has not measured and reported enterprise architecture outcomes and benefits. Though the department's 2008 Enterprise Architecture Performance Measurement Guide states that the department planned to collect and analyze data and to measure and report results on a quarterly basis starting the first quarter of calendar year 2009, it has yet to do so. In addition, while the department reported in its response to our survey on enterprise architecture results and outcomes that its architecture program contributed to an estimated $83 million in cost savings in fiscal year 2009, it did not provide documentation to support the cost savings estimate or evidence that the outcome was reported to agency executives.

Source: GAO analysis of agency-provided data.

Department of the Treasury

Table 20 shows the Department of the Treasury's satisfaction of relevant framework elements in version 2.0 of GAO's EAMMF.

Table 20: Department of the Treasury Satisfaction of EAMMF Elements

Element	Satisfied?	Summary
The enterprise architecture's intended purpose or strategic goals are defined.	●	Treasury has defined its enterprise architecture goals. Specifically, according to Treasury's E-Government Act Report for fiscal year 2011, its enterprise architecture plans have focused on reducing duplication through its data center consolidation initiative because infrastructure reflects the majority of the department's IT spending. Accordingly, Treasury has defined goals related to data center consolidation. Specifically, its goals include increased cost efficiency through consolidation of facilities and infrastructure, increased economies of scale and associated buying power, and reduced overhead associated with operating multiple instances of common facilities and services. According to officials, the department is in the process of developing broader goals that will integrate its capital planning and investment control and architecture processes.
A method and metrics have been established to measure enterprise architecture strategic mission value (outcomes and benefits).	◐	Treasury has established metrics for measuring enterprise architecture outcomes and benefits but has yet to document a method. Specifically, the department plans to measure a decrease in the number of servers, an increase in the percentage of operating systems that are virtual, and a decrease in demand for data center square footage. According to the department's September 2011 Data Center Consolidation Plan, reductions in physical assets should produce increases in capacity and cost efficiencies for management of space and IT services. However, the department has not established a methodology for measuring its architecture outcomes with detailed steps to be followed (including sources of information). Further, the Chief Enterprise Architect stated that metrics corresponding to broader goals that integrate capital planning and investment control and enterprise architecture are under development.
Enterprise architecture outcomes and benefits are periodically measured and reported to the agency's enterprise architecture executive committee.	◐	The department has measured and reported architecture outcomes associated with its data center consolidation. Specifically, its September 2011 Data Center Consolidation Plan, which was approved by the department's Deputy Assistant Secretary for Information Systems and Chief Information Officer, reported a reduction in the number of servers, an increase in the percentage of operating systems that were virtualized, and a reduction in data center square footage between 2010 and 2011. However, this metric has yet to be periodically measured and reported as an architecture outcome. As noted above, going forward, Treasury is developing new goals and metrics.

Source: GAO analysis of agency-provided data.

Department of Veterans Affairs

Table 21 shows the Department of Veterans Affairs' (VA) satisfaction of relevant framework elements in version 2.0 of GAO's EAMMF.

Table 21: Department of Veterans Affairs Satisfaction of EAMMF Elements

Element	Satisfied?	Summary
The enterprise architecture's intended purpose or strategic goals are defined.	◑	According to the Chief Architect, VA is in the process of developing an enterprise architecture program overview statement and guiding enterprise architecture principles. Specifically, according to draft documentation, VA's enterprise architecture is to provide tools, rules, principles, and standards to guide efficient, effective, and interoperable implementation of the department's vision of providing seamless delivery of benefits and services to veterans. Global Enterprise Architecture principles include that all VA solutions are to utilize enterprise-wide standards, services, and approaches to deliver seamless capabilities to veterans, facilitate IT consolidations through reuse, and simplify the use of VA functions. According to department officials, enterprise architecture principles have been finalized, but are not planned to be formally released until September 30, 2012.
A method and metrics have been established to measure enterprise architecture strategic mission value (outcomes and benefits).	○	According to the department's Chief Architect, VA has not established a method and metrics for measuring enterprise architecture outcomes and benefits because the department's approach to enterprise architecture and enterprise architecture governance is being revised. The Chief Architect said that once the approach is updated, an enterprise architecture value measurement plan will be developed. However, the official noted that it is a challenge to know how much to attribute outcomes to enterprise architecture relative to other factors in the decision-making process.
Enterprise architecture outcomes and benefits are periodically measured and reported to the agency's enterprise architecture executive committee.	○	VA has not periodically measured and reported enterprise architecture outcomes and benefits.

Source: GAO analysis of agency-provided data.

Environmental Protection Agency

Table 22 shows the Environmental Protection Agency's (EPA) satisfaction of relevant framework elements in version 2.0 of GAO's EAMMF.

Table 22: Environmental Protection Agency Satisfaction of EAMMF Elements

Element	Satisfied?	Summary
The enterprise architecture's intended purpose or strategic goals are defined.	●	EPA has defined its architecture goals and objectives. Specifically, according to the agency's February 2011 Modernization Blueprint, a major goal is to use the architecture to identify segments within the organization that could serve as candidates for service sharing and reuse. Additional goals and objectives are described in the agency's Office of Technology Operation and Planning Mission Investments Solution Division April 2011 draft strategic plan for fiscal years 2011 to 2016. These include, among other things, providing architecture services to enable stakeholders to mature and increase value from their architectures, and developing a common standard to be used in evaluating segments and solutions across the enterprise.
A method and metrics have been established to measure enterprise architecture strategic mission value (outcomes and benefits).	○	The agency has not established a method and metrics for measuring enterprise architecture outcomes and benefits. Specifically, agency officials reported that the results from an enterprise architecture management maturity self-assessment, completed in April 2012, will be analyzed to determine areas for improvement and reflected in the agency's enterprise architecture performance measurement plan. According to agency officials, the enterprise architecture program performance measurement plan is expected to be completed in fiscal year 2013.
		The agency has included metrics in its Office of Technology Operation and Planning Mission Investments Solution Division draft strategic plan and Enterprise Architecture Value Measures project charter approved in May 2012. These metrics include the percentage of segments and solution architectures that have been reviewed by the enterprise architecture program, the percentage of investments that identify future use of enterprise services, and the percentage of complete mandatory data fields in the agency's enterprise architecture repository. However, these metrics measure outputs (i.e., direct products and services) of the program rather than outcomes (i.e., results of enterprise architecture products and services such as benefits to Congress and the American taxpayer). According to officials, the metrics to measure cost savings and efficiencies from enterprise architecture will be identified in fiscal years 2013 and 2014. Officials also added that it has been challenging to measure performance because baselines have changed from year to year due to changing OMB reporting requirements.
Enterprise architecture outcomes and benefits are periodically measured and reported to the agency's enterprise architecture executive committee.	○	The agency is not measuring and reporting architecture outcomes and benefits.

Source: GAO analysis of agency-provided data.

General Services Administration

Table 23 shows the General Services Administration's (GSA) satisfaction of relevant framework elements in version 2.0 of GAO's EAMMF.

Table 23: General Services Administration Satisfaction of EAMMF Elements

Element	Satisfied?	Summary
The enterprise architecture's intended purpose or strategic goals are defined.	●	According to GSA's March 2011 briefing to OMB on its enterprise architecture modernization plan, its enterprise architecture provides services to increase interoperability between systems; increase reuse of systems, information, and services; increase agility and flexibility in building and operating systems; and facilitate achievement of agency goals. Further, according to GSA's March 2012 Enterprise Modernization Roadmap, enterprise architecture is to, among other things, increase system interoperability and cost efficiencies, reduce duplication, and increase innovation.
A method and metrics have been established to measure enterprise architecture strategic mission value (outcomes and benefits).	◐	GSA has established metrics but not a method for measuring enterprise architecture outcomes and benefits. Specifically, GSA's March 2012 Enterprise Modernization Roadmap includes enterprise architecture program work plans and corresponding output and outcome metrics. For example, GSA has identified the percentage of applications complying with IT standards as a metric for measuring the extent to which the agency is increasing its use of IT standards, which is one of its desired outcomes. The roadmap also includes a desired outcome of increasing development, modernization, and enhancement spending by 25 percent per year beginning in fiscal year 2014. According to the agency, the ratio of development, modernization, and enhancement to steady-state spending allows GSA's enterprise architecture program to highlight the allocation of IT spending and opportunities to reduce operating costs.
		However, GSA has not established a method for measuring and reporting enterprise architecture outcomes and benefits. According to GSA officials, the benefits of architecture are achieved early in system development and are difficult to relate to future return on investment measured later in a system's life cycle. Nonetheless, officials said they were beginning to develop a method for measuring enterprise architecture outcomes, but did not expect the plan to be completed for 2 to 5 years.
Enterprise architecture outcomes and benefits are periodically measured and reported to the agency's enterprise architecture executive committee.	○	The agency has not measured and reported enterprise architecture outcomes and benefits.

Source: GAO analysis of agency-provided data.

National Aeronautics and Space Administration

Table 24 shows the National Aeronautics and Space Administration's (NASA) satisfaction of relevant framework elements in version 2.0 of GAO's EAMMF.

Table 24: National Aeronautics and Space Administration Satisfaction of EAMMF Elements

Element	Satisfied?	Summary
The enterprise architecture's intended purpose or strategic goals are defined.	●	NASA's November 2011 Enterprise Architecture Policy identifies the purposes of the agency's enterprise architecture, which include • being a composition of architectures and set of integrated reference models that map all IT initiatives, capabilities, and services to agency needs; • serving to guide executive decision making, establishing a clear linkage between present capabilities and future NASA mission needs, including identifying potential shortfalls and redundancies in IT capabilities, the time frame in which the shortfall or redundancy exists, and an analysis of industry alternatives and remedial solutions/approaches; • providing a foundation for further development, modernization or modification, and enhancements of integrated architectures; • identifying mission IT dependencies; • being used as a tool to integrate strategic planning efforts and to select, guide, manage, rationalize, and prioritize agency investments; • establishing the framework for agency interoperability by providing the standard, rigorous construct for horizontal and vertical integration of mission needs and business processes through architecture; • being integral to the budget life cycle, enabling informed and timely procurement decision making to influence capital and strategic sourcing investments; and • promoting transparency and accountability by aligning functions/capabilities, services, systems, components, and related standards to agency strategy.
A method and metrics have been established to measure enterprise architecture strategic mission value (outcomes and benefits).	○	NASA has yet to establish a method or metrics for measuring enterprise architecture outcomes. However, according to NASA's Chief Enterprise Architect, an approach for measuring enterprise architecture performance is being developed. Specifically, draft enterprise architecture procedural requirements include metrics to measure the number of approved architecture artifacts. However, these metrics measure output (i.e., direct products and services) of the enterprise architecture program, rather than outcomes (i.e., results of enterprise architecture products and services such as benefits to Congress and the American taxpayer). Moreover, according to the Chief Architect, the agency does not yet have a mature enterprise architecture program and establishing one is a challenge because the agency's IT environment is not structured in a way that readily accepts an enterprise-wide architecture. Specifically, the official stated that much of NASA's funding is provided to the agency's centers, which invest the money to meet their specific needs with little regard for the agency's overall needs or existing capabilities. In July 2012, the NASA Chief Enterprise Architect stated that a NASA policy is expected to be issued by 2013, requiring that a method and metrics for measuring enterprise architecture value be established.
Enterprise architecture outcomes and benefits are periodically measured and reported to the agency's enterprise architecture executive committee.	○	NASA has yet to measure and report its architecture outcomes and benefits.

Source: GAO analysis of agency-provided data.

| National Science Foundation | Table 25 shows the National Science Foundation's (NSF) satisfaction of relevant framework elements in version 2.0 of GAO's EAMMF. |

Table 25: National Science Foundation Satisfaction of EAMMF Elements

Element	Satisfied?	Summary
The enterprise architecture's intended purpose or strategic goals are defined.	●	NSF has identified enterprise architecture goals in its September 2008 Information Resource Management Plan. These goals are: • Improve utilization of IT resources by eliminating duplicative investments, and promoting sharing of common services and standards. • Improve program performance by ensuring business functions support strategic goals and priorities, data are optimized in support of the business, and applications and technology solutions are driven by business needs. • Simplify IT investment decisions by providing a line of sight from strategy to business function to technology, which enables decision makers to select investments that support NSF's core mission, and to identify duplicative or misaligned initiatives. • Reduce IT diversity and complexity within NSF by promoting standards and the sharing and reuse of common technologies. • Improve interoperability through the establishment of enterprise-wide standards that promote platform and vendor independence, enabling greater interoperability across disparate applications, both internal and external.
A method and metrics have been established to measure enterprise architecture strategic mission value (outcomes and benefits).	○	NSF has yet to establish a method and metrics to measure enterprise architecture strategic mission value. Specifically, the agency has established metrics for measuring the percent of IT investments that comply with the agency's transition strategy, the percent of IT projects that comply with the agency's Enterprise Architecture Modernization Roadmap, the percent of IT services associated with an appropriate segment architecture, and the percent of approved software and technical architectures fulfilling opportunities to reuse shared services and IT infrastructure (i.e., ensuring that solution architectures in development reuse common infrastructure components or develop components for future reuse where possible). However, these metrics measure output (i.e., direct products and services) of the enterprise architecture program, rather than outcomes (e.g., cost avoidance, improved mission performance from reengineered business processes and modernizing systems, or benefits to Congress and the American taxpayer). NSF officials stated that the agency is exploring opportunities to mature its process for measuring enterprise architecture outcomes, and plans to revise its Enterprise Architecture Program Management Plan to align with new OMB enterprise architecture guidance by the end of fiscal year 2012.
Enterprise architecture outcomes and benefits are periodically measured and reported to the agency's enterprise architecture executive committee.	○	NSF has yet to measure and report enterprise architecture outcomes and benefits.

Source: GAO analysis of agency-provided data.

Nuclear Regulatory Commission

Table 26 shows the Nuclear Regulatory Commission's (NRC) satisfaction of relevant framework elements in version 2.0 of GAO's EAMMF.

Table 26: Nuclear Regulatory Commission Satisfaction of EAMMF Elements

Element	Satisfied?	Summary
The enterprise architecture's intended purpose or strategic goals are defined.	●	NRC has defined the purpose and goals for its architecture. Specifically, the purpose is to support IT goals that were established in its IT strategic plan for fiscal years 2012 through 2016. These include: • NRC staff and stakeholders can quickly and easily access the information they need. • IT business solutions are easy to use, cost effective, and strengthen agency performance, which according to agency officials, is focused on avoiding duplication and cost savings. • IT infrastructure is available, cost effective, and responsive to agency needs.
A method and metrics have been established to measure enterprise architecture strategic mission value (outcomes and benefits).	◑	NRC has established a metric for measuring enterprise architecture outcomes, but has yet to establish a method. Specifically, for the goal that the agency's IT infrastructure is available, cost effective, and responsive to agency business needs, NRC plans to measure progress toward having common access controls by measuring the reduction in passwords and/or sign-ons. However, a methodology with detailed steps for measuring enterprise architecture outcomes has not yet been established.
Enterprise architecture outcomes and benefits are periodically measured and reported to the agency's enterprise architecture executive committee.	○	NRC has not periodically measured and reported enterprise architecture outcomes.

Source: GAO analysis of agency-provided data.

Office of Personnel Management

Table 27 shows the Office of Personnel Management's (OPM) satisfaction of relevant framework elements in version 2.0 of GAO's EAMMF.

Table 27: Office of Personnel Management Satisfaction of EAMMF Elements

Element	Satisfied?	Summary
The enterprise architecture's intended purpose or strategic goals are defined.	●	According to OPM's IT Strategic Plan for 2010-2013, the agency's enterprise architecture defines IT management principals, goals, and objectives and establishes a roadmap to achieve the enterprise architecture vision of centralizing and managing OPM's IT infrastructure for the benefits and efficiencies that can be realized through technology. The IT strategic plan also includes the objective to utilize the enterprise architecture as a management and governance tool to strengthen decision making and standard setting, coordinating with OPM business lines to ensure technology decisions and implementations for new systems align with the agency's as well as the federal government's enterprise architecture.
A method and metrics have been established to measure enterprise architecture strategic mission value (outcomes and benefits).	◑	OPM has established cost savings as a metric to measure enterprise architecture results and outcomes, and developed an Enterprise Architecture Return on Investment Framework. According to the framework, return on investment is calculated over a period of time and relates the value contributed in dollars to the cost in dollars of the enterprise achitecture program. The framework identifies steps the enterprise architecture office plans to follow to determine the architecture's role in cost savings or improving mission, including defining enterprise architecture's role (strategic partner, collaborator, change agent etc.) in improving business and IT and the percentage to attribute to enterprise architecture for each role; collaboratively identifying the role enterprise architecture is supposed to play before each major effort; evaluating the actual role of enterprise architecture after each major effort; and soliciting feedback on how well the role was performed. However, the framework does not include steps to determine the cost savings or mission improvement to which enterprise architecture contributes or to calculate in dollars architecture's return on investment. According to OPM officials, one of the agency's challenges in developing a method is quantifying value based on the contributions of enterprise architecture to business/mission improvement.
Enterprise architecture outcomes and benefits are periodically measured and reported to the agency's enterprise architecture executive committee.	○	According to OPM officials, the agency has measured enterprise architecture cost savings and reported them to the Chief Information Officer. However, officials have not provided documentation to support that the cost savings have been reliably measured.

Source: GAO analysis of agency-provided data.

Small Business Administration

Table 28 shows the Small Business Administration's (SBA) satisfaction of relevant framework elements in version 2.0 of GAO's EAMMF.

Table 28: Small Business Administration Satisfaction of EAMMF Elements

Element	Satisfied?	Summary
The enterprise architecture's intended purpose or strategic goals are defined.	●	According to SBA's 2009 Capital Planning and Investment Control Policy Guide, the agency's enterprise architecture process is a management practice that is to support strategic planning; capital planning and investment control; system development; and IT asset management activities to optimize the agency's resources and achieve its performance goals. Specifically, according to the guide, decision makers are to leverage the agency's architecture to help ensure that investments • support the business needs, • address specific and measurable performance gaps, • align with the agency's mission and goals, • comply with the agency's standards, and • reduce or eliminate spending on unneeded, redundant, and/or duplicative IT assets.
A method and metrics have been established to measure enterprise architecture strategic mission value (outcomes and benefits).	○	SBA has not established a method or metrics to measure enterprise architecture outcomes and benefits. Nonetheless, according to agency officials, SBA measures the extent to which proposed IT investments align with the enterprise architecture during its capital planning and investment control process. However, such a metric measures output (i.e., direct products and services) of the program rather than outcomes (i.e., results of enterprise architecture products and services such as benefits to Congress and the American taxpayer) of the program. Agency officials also stated that the architecture program maintained Financial Assistance and Disaster Assistance segment architectures, which were used to facilitate planning and decision making, and that the agency achieved high- level performance goals in fiscal year 2011 associated with the segments. However, the agency did not provide documentation showing that a method and metrics had been established for measuring architecture outcomes, or that the high-level performance outcomes were linked to enterprise architecture.
Enterprise architecture outcomes and benefits are periodically measured and reported to the agency's enterprise architecture executive committee.	○	SBA is not periodically measuring enterprise architecture outcomes and benefits.

Source: GAO analysis of agency-provided data.

Social Security Administration

Table 29 shows the Social Security Administration's (SSA) satisfaction of relevant framework elements in version 2.0 of GAO's EAMMF.

Table 29: Social Security Administration Satisfaction of EAMMF Elements

Element	Satisfied?	Summary
The enterprise architecture's intended purpose or strategic goals are defined.	●	SSA has defined its enterprise architecture goals in its 2010 Enterprise Architecture Program Plan. Specifically, the architecture is to provide visibility for IT initiatives and support alignment with SSA's strategic business plans; support design and configuration management decisions and alignment of IT initiatives with SSA's infrastructure; and support decisions regarding operations, maintenance, and the development of IT resources and services.
		In addition, as we previously reported, SSA's enterprise architecture for years 2011 through 2016 described a vision that includes eliminating existing stove-piped application software, and reusing services to develop service-oriented architecture applications to replace aging online and back-office desktop applications.[a] According to the SSA's Enterprise Architecture Transition Strategy, these efforts are expected to help reduce costs and increase productivity.
A method and metrics have been established to measure enterprise architecture strategic mission value (outcomes and benefits).	○	SSA has yet to establish a method and metrics to measure enterprise architecture outcomes. SSA has established a metric to measure the number of IT projects compliant with agency architecture standards. In addition, SSA officials stated that they measure the percent of IT investments aligned to the agency's strategic portfolios. However, these metrics measure outputs (i.e., direct products and services) of the program, rather than outcomes (e.g., benefits to Congress and the American taxpayer). SSA officials stated that they are considering developing additional metrics to measure enterprise architecture value, for example, a metric to measure the extent to which architecture helps identify opportunities to reuse services and related software modules. However, they noted that the lack of guidelines and best practices contribute to the difficulty in measuring outcomes. To address this challenge, SSA officials said that they will participate in the Enterprise Architecture Value Measurement workgroup of the federal CIO Council Strategy and Planning Committee's Architecture Subcommittee. They added that as new metrics are identified they will be documented and communicated throughout the agency.
Enterprise architecture outcomes and benefits are periodically measured and reported to the agency's enterprise architecture executive committee.	○	SSA has not periodically measured and reported enterprise architecture outcomes and benefits.

Source: GAO analysis of agency-provided data.

[a]GAO, *Social Security Administration: Improved Planning and Performance Measures are Needed to Help Ensure Successful Technology Modernization*, GAO-12-495 (Washington, D.C.: Apr. 26, 2012).

GAO-12-791 Organizational Transformation

U.S. Agency for International Development

Table 30 shows the U.S. Agency for International Development's (USAID) satisfaction of relevant framework elements in version 2.0 of GAO's EAMMF.

Table 30: U.S. Agency for International Development Satisfaction of EAMMF Elements

Element	Satisfied?	Summary
The enterprise architecture's intended purpose or strategic goals are defined.	●	USAID has developed architecture goals. Specifically, the agency's December 2011 Enterprise Architecture Program Charter identifies the following goals: <ul><li>Support improvement of mission-critical business processes through business process analysis, and identification and application of enterprise architecture standards.</li><li>Guide analytical efforts to locate, validate, and promote the strategic use of agency information.</li><li>Facilitate analysis of the agency's IT environment, including IT hardware, software, and enterprise applications, to promote the effective and efficient deployment of IT services.</li><li>Provide governance for USAID technology efforts by designing and supporting the implementation of enterprise architecture models and standards.</li></ul>
A method and metrics have been established to measure enterprise architecture strategic mission value (outcomes and benefit).	●	USAID has established two metrics for measuring enterprise architecture outcomes: (1) cost savings and avoidance due to process efficiency, technology standardization, retirement, and consolidation; and (2) client satisfaction based on client survey responses on architecture's value to business (such as facilitating decision making on technology and processes using enterprise architecture tools). The agency has also established guidance for measuring cost savings and avoidance and an approach to measure client satisfaction by developing and implementing surveys.
Enterprise architecture outcomes and benefits are periodically measured and reported to the agency's enterprise architecture executive committee.	●	The agency has periodically measured enterprise architecture outcomes and, according to agency officials, these outcomes are reported to the Deputy CIO and CIO. In addition, the agency has established a website for CIO staff, including the Deputy CIO and CIO, to review monthly architecture outcomes. According to its February 2012 Enterprise Architecture Performance Results report, the agency achieved $12.3 million in savings and $9.5 million in cost avoidance by transitioning disparate human resource systems to a human resource shared services center. The Federal Enterprise Architecture and the agency's enterprise architecture were used to select a shared services center. In addition, the agency reported estimated savings of $15.7 million (not including $4 million in migration costs) over the next 5 years, beginning in fiscal year 2013, by moving its e-mail service to a cloud-based service. According to agency officials, the cloud-based solution was recommended by the architecture team because it can replace multiple installations of the current e-mail solution. According to the agency's return on investment analysis, this will reduce hardware and software maintenance, and labor and other expenses. However, as of September 2012, the new service had yet to be approved. Officials explained that delay in approval was causing a reduction in the return on investment and that the cost savings were being updated, accordingly.

Source: GAO analysis of agency-provided data.

Appendix III: Comments from the Department of Labor

U.S. Department of Labor

Office of the Assistant Secretary
for Administration and Management
Washington, D.C. 20210

AUG 2 9 2012

Ms. Valerie C. Melvin
Director
Information Management and Technology Resources Issues
Government Accountability Office
441 G St. NW
Washington, D.C. 20548

Dear Ms. Melvin:

Thank you for the opportunity to review and comment on the Draft Government Accountability Office (GAO) Report # GAO-12-791, *Organizational Transformation: Enterprise Architecture Value Needs to be Measured and Reported.* We appreciate the GAO's efforts and the insight provided by the report.

After carefully reviewing the draft GAO report, the Department of Labor has no comments to contribute at this time.

Should you have any questions regarding the Department's response, please contact Mr. Thomas Markey, Acting Deputy Chief Information Officer, at markey.tom@dol.gov or 202-693-4220.

Sincerely,

T. Michael Kerr
Assistant Secretary for
Administration and Management

cc: Thomas Markey, Acting Deputy Chief Information Officer

Appendix IV: Comments from the Department of the Treasury

DEPARTMENT OF THE TREASURY
WASHINGTON, D.C. 20220

SEP – 5 2012

Neelaxi Lakhmani
Assistant Director
Information Management and Technology Resource Issues
U.S. Government Accountability Office
441 G Street, NW
Washington, DC 20548

Dear Ms. Lakhmani,

Thank you for the opportunity to provide comments on GAO's Draft Report, *"Organizational Transformation: Enterprise Architecture Value Needs to be Measured and Reported (GA0-12-791)."* The Department of Treasury has no comments on the Report and appreciates GAO's efforts in its development.

Please contact me at 202-622-1200 if you need anything further.

Sincerely,

Robyn East
Deputy Assistant Secretary for Information Systems
and Chief Information Officer

Appendix V: Comments from the Department of Agriculture

United States
Department of
Agriculture

Office of the Chief
Information Officer

1400 Independence
Avenue SW

Washington, DC
20250

SEP 0 5 2012

Valarie Melvin
Director, Information Management and Technology Resources Issues
Government Accountability Office
U.S. Government Accountability Office
441 G Street, N. W.
Washington, DC 20548

Dear Ms. Melvin,

The Department of Agriculture appreciates the opportunity to review and comment on the GAO Draft Report GAO-12-791 (Organizational Transformation: Enterprise Architecture Value Needs to be Measured and Reported, dated September 2012 (#310964)).

USDA concurs with the GAO findings and recommendations.
Again, thank you for the opportunity to review the draft report. If you have any questions, contact me at (202) 720- 8833.

Sincerely,

Cheryl L. Cook
Acting Chief Information Officer

General Comments of the USDA on GAO Draft Report GAO-12-791 (Organizational Transformation: Enterprise Architecture Value Needs to be Measured and Reported, dated September 2012

The Department appreciates the opportunity to review and comment on the Draft Report.

GAO Summary for Recommendations:

We also recommend that the Secretaries of the Departments of Agriculture, the Air Force, the Army, Commerce, Defense, Education, Energy, Homeland Security, the Interior, Labor, the Navy, State, Transportation, the Treasury, and Veterans Affairs; the Attorney General; the Administrators of the Environmental Protection Agency, General Services Administration, National Aeronautics and Space Administration, and Small Business Administration; the Commissioners of the Nuclear Regulatory Commission and Social Security Administration; and the Directors of the National Science Foundation and the Office of Personnel Management ensure the following two actions are taken:

- fully establish an approach for measuring enterprise architecture outcomes, including a documented method (i.e., steps to be followed) and metrics that are measurable, meaningful, repeatable, consistent, actionable, and aligned with the agency's enterprise architecture's strategic goals and intended purpose; and

- periodically measure and report enterprise architecture outcomes and benefits to top agency officials (i.e., executives with authority to commit resources or make changes to the program) and to OMB.

USDA's Response:

In 2010, OMB hired a Federal Chief Enterprise Architect to assist OMB in improving the Federal Architecture posture. The Federal Chief Architect established working groups to revise the Federal Enterprise Architecture Framework, with the goal of creating a "Common Approach to Federal Enterprise Architecture". In May 2012, OMB released the *Common Approach to Federal Enterprise Architecture* to promote increased levels of mission effectiveness by standardizing the development and use of architectures within and between federal agencies. OMB has not yet provided sufficient details on the method and metrics that could be used to measure architecture program outcomes. The Federal Chief Enterprise Architect plans to provide agencies more detailed guidance on measuring enterprise architecture as part of the Collaborative Planning Methodology (CPM), which is scheduled to be completed in FY13. Once the OMB guidance is provided, USDA will develop metrics and guidance to comply with the Federal Chief Architect's guidance.

Appendix VI: Comments from the Department of Commerce

UNITED STATES DEPARTMENT OF COMMERCE
The Secretary of Commerce
Washington, D.C. 20230

September 6, 2012

Ms. Valerie C. Melvin
Director, Information Management and
    Technology Resources Issues
U.S. Government Accountability Office
Washington, DC 20548

Dear Ms. Melvin:

Thank you for the opportunity to comment on the draft report from the U.S. Government Accountability Office (GAO) entitled *Organizational Transformation: Enterprise Architecture Value Needs to be Measured and Reported* (GAO-12-791).

We agree with the general findings as they apply to the Department of Commerce and with the specific recommendations relating to Commerce's measuring and reporting of enterprise architecture outcomes.

We developed a means of measuring enterprise architecture outcomes through the metrics defined in the Balanced Scorecard process. Our Enterprise Architecture Balanced Scorecard is one measure in a Department-wide performance measurement program. We are now completing the first full year of assessing enterprise architecture performance and will analyze both the assessment process and the associated findings to make improvements for the coming year.

Please contact Jerry Harper, Acting Director, Office of IT Policy and Planning, at 202-482-0222 if you have questions.

Sincerely,

Rebecca M. Blank
Acting Secretary of Commerce

Appendix VII: Comments from the Department of Defense

DEPARTMENT OF DEFENSE
6000 DEFENSE PENTAGON
WASHINGTON, D.C. 20301-6000

CHIEF INFORMATION OFFICER

Ms. Valerie C. Melvin
Director, Information Technology Management and Technology Resources Issues
U.S. Government Accountability Office
441 G Street, N.W.
Washington, DC 20548

Dear Ms. Melvin:

This is the Department of Defense (DoD) response to the GAO Draft Report, GAO-12-791, "ORGANIZATIONAL TRANSFORMATION: Enterprise Architecture Value Needs to be Measured and Reported," dated September 2012 (GAO Code 310964).

The Department appreciates the opportunity to comment. We concur with GAO recommendations. Our specific response to each recommendation is provided.

Sincerely,

David L. DeVries
Deputy Chief Information Officer for
Information

Enclosure:
As stated

GAO DRAFT REPORT DATED SEPTEMBER, 2012

GA0-12-791 (GAO CODE 310964)

"ORGANIZATIONAL TRANSFORMATION: ENTERPRISE ARCHITECTURE

VALUE NEEDS TO BE MEASURED AND REPORTED"

DEPARTMENT OF DEFENSE COMMENTS

TO THE GAO RECOMMENDATIONS

RECOMMENDATION 1: Fully establish an approach for measuring enterprise architecture outcomes, including a documented method (i.e., steps to be followed) and metrics that are measurable, meaningful, repeatable, consistent, actionable, and aligned with the agency's enterprise architecture's strategic goals and intended purpose.

DoD RESPONSE: The DoD concurs The DoD CIO is developing an Enterprise Architecture (EA) Management Plan (EAMP) that provides high level processes, to include measuring EA outcomes, in addition to policy in the form of a DoD Instruction.

RECOMMENDATION 2: Periodically measure and report enterprise architecture outcomes and benefits to top agency officials (i.e., executives with authority to commit resources or make changes to the program) and to OMB.

DoD RESPONSE: The DoD concurs. The DoD CIO staff is incorporating metrics into the yearly Enterprise Roadmap required by OMB, as well as developing measures for inclusion in the Department's Strategic Management Plan for business operations, and the Department's performance reporting for the Government Performance Results Act. The requirement for metrics will be further strengthened by inclusion in the DoD Instruction, with procedures for reporting and collection documented in the EAMP.

Appendix VIII: Comments from the Department of Education

UNITED STATES DEPARTMENT OF EDUCATION

OFFICE OF THE CHIEF INFORMATION OFFICER

THE CHIEF INFORMATION OFFICER

September 6, 2012

Ms. Valerie Melvin
Director
Information Management and Technology Resources Issues
Government Accountability Office
441 G Street, NW
Washington, DC 20548

Dear Ms. Melvin:

I am writing to respond to recommendations made in the Government Accountability Office (GAO) draft report, GAO-12-791, "Organizational Transformation: Enterprise Architecture Value Needs to be Measured and Reported." This report focused on improving the measurement and reporting of enterprise architecture (EA) outcomes and benefits.

The U.S. Department of Education (Department) appreciates the opportunity to respond to the GAO report and the need for improved measurement and reporting of EA outcomes and benefits.

Our responses to GAO's specific recommendations for executive action to the Secretary of Education regarding the measurement and reporting of EA outcomes and benefits follow.

Recommendation 1: *Fully establish an approach for measuring enterprise architecture outcomes, including a documented method (i.e., steps to be followed) and metrics that are measurable, meaningful, repeatable, consistent, actionable, and aligned with the agency's enterprise architecture's strategic goals and intended purpose.*

Response: The Department concurs with this recommendation. The Department has included a number of EA program performance metrics in its current Enterprise Roadmap submission (August 31, 2012). We intend to strengthen our existing metrics and develop additional metrics that express the EA program's goals for the Department. We will develop, document, and implement a measurement and reporting method that will be used to periodically monitor our progress toward achieving the goals, desired outcomes, and benefits described in our enterprise architecture program. In conjunction with OMB guidance, we will continue to use the Enterprise Roadmap as one of the mechanisms for reporting EA program performance to agency executive leadership.

Recommendation 2: *Periodically measure and report enterprise architecture outcomes and benefits to top agency officials (i.e., executives with authority to commit resources or make changes to the program) and to OMB.*

400 MARYLAND AVE. S.W., WASHINGTON, DC 20202
www.ed.gov

The Department of Education's mission is to promote student achievement and preparation for global competitiveness by fostering educational excellence and ensuring equal access.

Response: The Department concurs with this recommendation. As part of our enterprise architecture program, we will periodically report enterprise architecture outcomes and benefits to the Investment Review Board. When the Federal Chief Enterprise Architect releases the promised guidance (pg. 26 of the report) on measuring EA value, we will comply with that guidance.

Again, I appreciate the opportunity to respond to the GAO report. If you or your staff have any questions regarding our response, please contact me at (202) 245-6252 or Ken Moore at (202) 245-6908 or ken.moore@ed.gov.

Sincerely,

Danny A. Harris, Ph.D.

2

Appendix IX: Comments from the Department of Homeland Security

U.S. Department of Homeland Security
Washington, DC 20528

Homeland Security

September 19, 2012

Valerie C. Melvin
Director, Information Management and Technology Resources Issues
U.S. Government Accountability Office
441 G Street, NW
Washington, DC 20548

Re: Draft Report GAO-12-791, "ORGANIZATIONAL TRANSFORMATION:
        Enterprise Architecture Value Needs to be Measured and Reported"

Dear Ms. Melvin:

Thank you for the opportunity to review and comment on this draft report. The U.S. Department of Homeland Security (DHS) appreciates the U.S. Government Accountability Office's (GAO's) work in planning and conducting its review and issuing this report.

The Department is pleased to note GAO's positive acknowledgement that the intended purpose and strategic goals for its enterprise architecture are fully defined within the Enterprise Architecture Strategic Plan for Fiscal Years 2012-2016.

The draft report contained two recommendations with which the Department concurs. Specifically, GAO recommended that the Secretary of Homeland Security:

Recommendation 1: Fully establish an approach for measuring enterprise architecture outcomes, including a documented method (i.e., steps to be followed) and metrics that are measurable, meaningful, repeatable, consistent, actionable, and aligned with the agency's enterprise architecture's strategic goals and intended purpose.

Response: Concur. DHS will continue to strengthen its methodology for measuring and documenting enterprise architecture outcomes and metrics. Specifically, DHS will leverage existing enterprise architecture scoring mechanisms such as the Office of Management and Budget's (OMB's) Exhibits 53 (Information Technology and E-Government) and 300 (Planning, Budgeting, Acquisition, and Management of Information Technology Capital Assets) alignment scoring criteria, and the Architecture Maturity criteria for evaluating each major information technology investments' architecture maturity within the DHS Quarterly Program Accountability Report, along with the Data Architecture Management scorecard which identifies cost avoidance for the enterprise architecture process.

Recommendation 2: Periodically measure and report enterprise architecture outcomes and benefits to top agency officials (i.e., executives with authority to commit resources or make changes to the program) and to OMB.

Response: Concur. DHS will brief enterprise architecture outcomes for goals and objectives outlined in the Enterprise Architecture Strategic Plan to the DHS Chief Information Officer Council by October 31, 2012. DHS will ensure that OMB guidance, to be issued by December 31, 2012, for enterprise architecture value measurement and reporting is incorporated into our method for measuring outcomes, and that the results of those measurements are included in the DHS annual Enterprise Roadmap submission to OMB.

Again, thank you for the opportunity to review and comment on this draft report. Please feel free to contact me if you have any questions. We look forward to working with you in the future.

Sincerely,

Jim H. Crumpacker
Director
Departmental GAO-OIG Liaison Office

2

Appendix X: Comments from the Department of the Interior

United States Department of the Interior

OFFICE OF THE SECRETARY
Washington, D.C. 20240

SEP 10 2012

Ms. Valerie C. Melvin
Director
Information Management
 and Technology Resources Issues
U.S. Government Accountability Office
441 G Street, N.W.
Washington, D.C. 20548

Dear Ms. Melvin:

Thank you for providing the Department of the Interior (DOI) the opportunity to review and comment on the draft Government Accountability Office Report entitled, *ORGANIZATIONAL TRANSFORMATION. Enterprise Architecture Value Needs to be Measured and Reported* (GAO-12-791).

DOI concurs with the following two recommended actions and has no additional comments.

- DOI will establish an approach for measuring enterprise architecture outcomes, including a documented method (i.e., steps to be followed) and metrics that are measurable, meaningful, repeatable, consistent, actionable, and aligned with DOI's enterprise architecture's strategic goals and intended purpose;
- DOI will periodically measure and report enterprise architecture outcomes and benefits to top agency officials (i.e., executives with authority to commit resources or make changes to the program) and to OMB.

If you have any questions or need additional information, please contact Kelly Morrison, Acting Director, Enterprise Architecture at (202) 208-5413.

Sincerely,

Rhea Suh
Assistant Secretary
Policy Management and Budget

Appendix XI: Comments from the Department of State

United States Department of State

Comptroller
1969 Dyess Avenue
Charleston, SC 29405

SEP 1 2 2012

Dr. Loren Yager
Managing Director
International Affairs and Trade
Government Accountability Office
441 G Street, N.W.
Washington, D.C. 20548-0001

Dear Dr. Yager:

We appreciate the opportunity to review your draft report, "ORGANIZATIONAL TRANSFORMATION: Enterprise Architecture Value Needs to be Measured and Reported" GAO Job Code 310964.

The enclosed Department of State comments are provided for incorporation with this letter as an appendix to the final report.

If you have any questions concerning this response, please contact Colleen Hinton, IT Manager, Bureau of Information Resource Management at (202) 634-0320.

Sincerely,

James L. Millette

cc: GAO – Valerie C. Melvin
     IRM – Steven Taylor (Acting)
     State/OIG – Evelyn Klemstine

U.S. Department of State Comments on GAO Draft Report

ORGANIZATIONAL TRANSFORMATION: Enterprise Architecture Value Needs to be Measured and Reported
(GAO-12-791, GAO Code 310964)

The U.S. Department of State has reviewed the GAO draft report entitled *"ORGANIZATIONAL TRANSFORMATION: Enterprise Architecture Value Needs to be Measured and Reported,"* and concurs, with comment, with the overall report's conclusions and recommendations.

GAO Comment:

"The Department has yet to establish a method or metrics for measuring enterprise architecture outcomes and benefits. Agency officials stated that while their objective is to have good IT investments, it is difficult to measure enterprise architecture's contribution because IT investment results are due to many factors, including good project management, and adequate funding, as well as enterprise architecture. While, according to the agency's IT Tactical Plan, a key performance indicator for its enterprise architecture program is evidence of increased use and value of enterprise architecture products and services in providing consistent and effective IT solutions, promoting interoperability, information sharing, and collaboration, State has yet to establish metrics and a method for measuring the value of its enterprise architecture products. Department officials reported that they expect to create metrics by December 2012."

GAO Recommendation: Fully establish an approach for measuring enterprise architecture outcomes, including a documented method (i.e., steps to be followed) and metrics that are measurable, meaningful, repeatable, consistent, actionable, and aligned with the agency's enterprise architecture's strategic goals and intended purpose.

U.S. Department of State Response:

The Department concurs with this recommendation. The Department continues to develop its Business, Technology, and Information Architectures, along with a Transition Plan that provides the foundation to

2

expand the IT Tactical Plan's use and value by establishing consistent and effective IT alignment, guidance, information awareness, and collaboration. The proposed Enterprise Technology Review Board will review proposed IT solutions to ensure their alignment to the Department's strategic IT goals and objectives

In addition, the Chief Information Officer (CIO) established a requirement for IT service lines, programs, and projects to align with the Department's IT Strategic and Tactical plans as a foundational methodology for evaluating IT investments and the EA framework.

The CIO also established a requirement to reduce the number of information exchange elements between enterprise systems creating an opportunity to streamline the Department's information flow, improve the quality of data sets, and system interoperability. The Enterprise Architecture office will facilitate this effort and measure the integration progress in partnership with the Department's Application and Data Coordination Working Group.

GAO Recommendation: Periodically measure and report enterprise architecture outcomes and benefits to top agency officials (i.e., executives with authority to commit resources or make changes to the program) and to OMB.

U.S. Department of State Response:

The Department concurs with this recommendation.

The specific metrics being proposed for measuring the use and value of EA are as follows:

1. Percentage of major IT investments aligned to the Department's IT Strategic Plan.
2. Percentage of major IT investments with a defined target architecture and technology roadmap.
3. Percentage of information exchange elements reduced between critical management systems.

To support the Department's established IT investment review process, the Department will utilize an EA Scorecard to monitor the progress of IT

3

investments, as they relate to decision making and the alignment of the Department's strategic goals, business processes, information needs, and enabling technology solutions. The above metrics will be implemented by the next IT investment review cycle, which is scheduled for the second and third quarters of FY 2013.

The Enterprise Architecture office will continue to collaborate with IT investment owners to produce useful EA products that strengthen IT programs, facilitate effective decision making and metrics for use in the evaluation process.

Appendix XII: Comments from the Department of Veterans Affairs

DEPARTMENT OF VETERANS AFFAIRS
Washington DC 20420

September 7, 2012

Ms. Valerie C. Melvin
Director, Information Technology
  Human Capital and Management Issues
U.S. Government Accountability Office
441 G Street, NW
Washington, DC 20548

Dear Ms. Melvin:

The Department of Veterans Affairs (VA) has reviewed the Government Accountability Office's (GAO) draft report, *"ORGANIZATIONAL TRANSFORMATION: Enterprise Architecture Value Needs to be Measured and Reported"* (GAO-12-791). VA generally agrees with GAO's conclusions and concurs with both recommendations to the Department.

The enclosure specifically addresses GAO's recommendations and provides comments to the draft report. VA appreciates the opportunity to comment on your draft report.

Sincerely,

John R. Gingrich
Chief of Staff

Enclosure

Department of Veterans Affairs (VA) Comments to
Government Accountability Office (GAO) Draft Report:
*"ORGANIZATIONAL TRANSFORMATION: Enterprise Architecture Value
Needs to be Measured and Reported"*
(GAO-12-791)

GAO Recommendation: We recommend that the Secretary of Veterans Affairs
ensure the following two actions are taken:

<u>Recommendation 1</u>: fully establish an approach for measuring enterprise
architecture outcomes, including a documented method (i.e., steps to be
followed) and metrics that are measurable, meaningful, repeatable, consistent,
actionable, and aligned with the agency's enterprise architecture's strategic goals
and intended purpose; and

<u>Recommendation 2</u>: periodically measure and report enterprise architecture
outcomes and benefits to top agency officials (i.e., executives with authority to
commit resources or make changes to the program) and to OMB.

VA Comment: Concur with both recommendations. The Department recognizes the
need to assess return on all agency investments, including Enterprise Architecture
(EA). Given the emerging state of most VA EA capabilities, however, no EA
performance measurement program is currently in place. An EA performance
management and measurement program is not possible until, 1) the Department's EA is
sufficiently robust and mature to guide and constrain development, and 2) use of the EA
to guide and constrain is effectively embedded within the Department's core processes.
VA is working very diligently along both tracks. It is VA's goal to have a formalized EA
measure program in place early in fiscal year (FY) 2014.

VA's EA program was essentially re-established in FY 2012 from the ground up. During
FY 2012, VA delivered:

- First ever enterprise-wide One-VA EA business architecture models.

- A series of technical EA products that define EA rules and standards for the
  Enterprise Technical Architecture (ETA) layer of the One-VA EA to which all
  programs must adhere.

- A consolidated compliance criteria document has been developed that guides users
  (both developers and investment managers) in how to apply ETA guidance
  throughout the investment lifecycle.

- Developed and published a future state architectural view of VA's technical IT
  environment, detailing how emerging technologies will influence VA's service
  offerings and internal processes going forward.

During FY 2013 VA will focus on continuing to mature the Department's EA and
embedding use of EA in VA core processes. VA will:

1

Enclosure

Department of Veterans Affairs (VA) Comments to
Government Accountability Office (GAO) Draft Report:
*"ORGANIZATIONAL TRANSFORMATION: Enterprise Architecture Value
Needs to be Measured and Reported"*
(GAO-12-791)

- Embed ETA compliance in the Department's program development/management milestone processes.   (1st Quarter FY 2013/2nd Quarter FY 2013)

- Partner with the Office of Planning and Policy to embed EA usage in the emerging consolidated VA requirements process.   (1st Quarter FY 2013/2nd Quarter FY 2013)

- Partner with VA administrations and VACO staff offices to expand VA business architecture capabilities.   (Throughout FY 2013)

- Partner with the VA Strategic Planning team to develop an integrated VA Enterprise Transition Plan.   (3rd Quarter FY 2013)

With these accomplishments VA will be in position to develop and implement a formal VA EA measurement program. During the second half of FY 2013, as VA gains experience in using EA in its application development/management processes, VA will be able to capture and report isolated examples of savings due to EA usage. Based on that experience, VA will use these examples to assist in developing, publishing and implementing a formal measurement program in 1st Quarter FY 2014.

2

National Aeronautics and Space Administration

Headquarters
Washington, DC 20546-0001

SEP 1 3 2012

Reply to Attn of:

Office of the Chief Information Officer

Ms. Valerie C. Melvin
Director
Information Management and Technology Resources Issues
United States Government Accountability Office
Washington, DC 20548

Dear Ms. Melvin:

The National Aeronautics and Space Administration (NASA) appreciates the opportunity to review and comment on the Government Accountability Office (GAO) draft report entitled, "ORGANIZATIONAL TRANSFORMATION: Enterprise Architecture Value Needs to be Measured and Reported" (GAO-12-791).

In the draft report, GAO addresses two recommendations to the NASA Administrator intended to enhance NASA's ability to realize enterprise architecture benefits. Specifically, GAO recommends that NASA:

Recommendation 1: Fully establish an approach for measuring enterprise architecture outcomes, including a documented method (i.e., steps to be followed) and metrics that are measurable, meaningful, repeatable, consistent, actionable, and aligned with the agency's enterprise architecture's strategic goals and intended purpose.

Management's Response: NASA concurs with GAO Recommendation 1. NASA is in the process of updating its NASA Procedural Requirements (NPR) for Enterprise Architecture (NPR 2830.1) that will address this recommendation. The NPR establishes Enterprise Architecture as a key element of Agency IT governance. The newly revised version of the NPR will better align enterprise architecture metrics and methods (i.e., processes) to measure outcomes (vs. output as cited in the GAO report) and benefits. Once the Office of Management and Budget (OMB) provides the "December 2012 Guidance for Enterprise Architecture Value Measurement and Reporting," NASA will evaluate its approach and make appropriate changes in the revised NPR. This action will be completed by June 2013.

Recommendation 2: Periodically measure and report enterprise architecture outcomes and benefits to top agency officials (i.e., executives with authority to commit resources or make changes to the program) and to OMB.

2

Management's Response: NASA concurs with GAO Recommendation 2. Once approved, the NPR 2830.1 processes will become the standard for measuring the enterprise architecture metrics and this information will be communicated to relevant Agency IT boards. The NPR incorporates Enterprise Architecture metrics into OMB reports as a matter of the standard business process. The forthcoming OMB requirements to report methods and metrics will be incorporated into these processes as appropriate. This action will be completed by June 2013.

Thank you for the opportunity to comment on this draft report. If you have any questions or require additional information, please contact John Hopkins at (202) 358-2519.

Sincerely,

Linda Cureton
Chief Information Officer

Appendix XIV: Comments from the Social Security Administration

SOCIAL SECURITY

Office of the Commissioner

September 7, 2012

Ms. Valerie C. Melvin
Director, Information Management and Technology Resources Issues
United States Government Accountability Office
441 G. Street, NW
Washington, D.C. 20548

Dear Ms. Melvin:

Thank you for the opportunity to review the draft report, "ORGANIZATIONAL TRANSFORMATION: Enterprise Architecture Value Needs to be Measured and Reported" (GAO-12-791). Our response is enclosed.

If you have any questions, please contact me at (410) 965-0520. Your staff may contact Amy Thompson, Senior Advisor for Audits, at (410) 966-0569.

Sincerely,

Dean S. Landis
Deputy Chief of Staff

Enclosure

SOCIAL SECURITY ADMINISTRATION   BALTIMORE, MD 21235-0001

COMMENTS ON THE GOVERNMENT ACCOUNTABILITY OFFICE DRAFT REPORT, "ORGANIZATIONAL TRANSFORMATION: ENTERPRISE ARCHITECTURE VALUE NEEDS TO BE MEASURED AND REPORTED" (GAO-12-791)

General Comment

The results of your report highlight the challenges inherent in measuring and reporting enterprise architecture (EA) outcomes and benefits. Only 3 agencies of the 27 surveyed have fully developed a method to measure EA mission outcomes, and only 1 agency -- the second smallest agency surveyed -- is currently measuring and reporting outcomes and benefits. The small number of agencies that are successfully measuring and reporting outcomes is an indication of the challenges inherent in attributing mission benefits to an EA program. We have not traditionally attributed many of the savings referenced in your report, such as consolidation of redundant systems and technology platforms, process efficiencies, use of virtualization in our data center, reuse of enterprise services, and consolidated use of government-wide procurement vehicles to our EA program. Rather, we have attributed the savings to specific investments. Every component in our information technology organization strives to achieve these outcomes.

Recommendation 1

Fully establish an approach for measuring enterprise architecture outcomes, including a documented method (i.e., steps to be followed) and metrics that are measureable, meaningful, repeatable, consistent, actionable, and aligned with the agency's enterprise architecture's strategic goals and intended purpose.

Response

We agree. As we develop and implement our approach for measuring EA outcomes and benefits, we will work to link our EA program outcomes to our strategic objectives.

Recommendation 2

Periodically measure and report enterprise architecture outcomes and benefits to top agency officials (i.e., executives with authority to commit resources or make changes to the program) and to OMB.

Response

We agree. We will compile the sort of measures described and provide them in a consolidated report of EA outcomes.

Appendix XV: Comments from the Environmental Protection Agency

UNITED STATES ENVIRONMENTAL PROTECTION AGENCY
WASHINGTON, D.C. 20460

SEP - 6 2012

OFFICE OF
ENVIRONMENTAL INFORMATION

Valerie C. Melvin, Director
Information Management and Technology Resources Issues
U.S. Government Accountability Office
441 G Street, NW
Washington, D.C. 20548

Dear Ms. Melvin:

Thank you for the opportunity to comment on the Draft GAO report entitled "Organizational Transformation: Enterprise Architecture Value Needs to be Measured and Reported (GAO-12-791)."

We agree with GAO's comment that EPA has defined its architecture goals, objectives, and output metrics but need to develop outcomes and benefits metrics. In recognition of GAO's comment, EPA plans to analyze the results of the enterprise architecture maturity self assessment to determine areas of improvement in performance management. Additionally, our review will include the Office of Management and Budget's (OMB) planned December 2012 guidance for enterprise architecture value measurement and reporting.

A performance measurement plan will be developed in FY 2013. The plan will identify processes to measure enterprise architecture outcomes and benefits. EPA will then periodically measure and report enterprise architecture outcomes and benefits to top Agency officials and to OMB in accordance with your report's recommendations for executive action.

Please feel free to contact Param Soni, Chief Architect in the Office of Environmental Information (OEI) at (202) 566-1177, if you would like to discuss these points any further.

Sincerely,

Malcolm D. Jackson
Assistant Administrator
and Chief Information Officer

Appendix XVI: Comments from the Department of Health and Human Services

DEPARTMENT OF HEALTH & HUMAN SERVICES

OFFICE OF THE SECRETARY

Assistant Secretary for Legislation
Washington, DC 20201

SEP 1 0 2012

Valerie C. Melvin, Director
Information Management and Technology Resources Issues
U.S. Government Accountability Office
441 G Street NW
Washington, DC 20548

Dear Ms. Melvin:

Attached are comments on the U.S. Government Accountability Office's (GAO) report entitled, "ORGANIZATIONAL TRANSFORMATION: Enterprise Architecture Value Needs to be Measured and Reported" (GAO-12-791).

The Department appreciates the opportunity to review this report prior to publication.

Sincerely,

Jim R. Esquea
Assistant Secretary for Legislation

Attachment

<u>**GENERAL COMMENTS OF THE DEPARTMENT OF HEALTH AND HUMAN
SERVICES (HHS) ON THE GOVERNMENT ACCOUNTABILITY OFFICE'S (GAO)
DRAFT REPORT ENTITLED, "ORGANIZATIONAL TRANSFORMATION:
ENTERPRISE ARCHITECTURE VALUE NEEDS TO BE MEASURED AND
REPORTED" (GAO-12-791)**</u>

The Department appreciates the opportunity to comment on this draft report. HHS concurs with
the overall findings of the draft report, and offers the following comments:

- OCIO is working closely with HHS Assistant Secretary for Financial Resources, to create
  new or revise existing value criteria to ensure alignment with Presidential initiatives and the
  directives in OMB memoranda and circulars, such as FY 2014 budget guidance to identify IT
  reductions and opportunities for reinvestments. HHS is also pursuing alternatives to improve
  and update enterprise architecture (EA) value measurements, related guidance, and reporting.
  Additionally, HHS is actively involved in the Federal workgroup on EA value measurements
  to help identify a standard methodology of reporting value measurement across the federal
  space.

1

Appendix XVII: Comments from the Department of Energy

Department of Energy
Washington, DC 20585

September 7, 2012

Ms. Valerie C. Melvin
Director, Information Management and
    Technology Resources Issues
U.S. Government Accountability Office
441 G Street, NW
Washington, D.C. 20548

Dear Ms. Melvin:

The Department of Energy's (DOE) Office of the Chief Information Officer (OCIO) appreciates the opportunity to comment on the General Accountability Office's (GAO) draft report, *Enterprise Architecture Value Needs to be Measured and Reported* (GAO-12-791).

DOE agrees that the effective use of Enterprise Architecture (EA) can be important to achieving operations and technology environments that maximize institutional mission performance and outcomes. Furthermore, we agree that utilizing enterprise architecture effectively will lead to cost savings through consolidation and the reuse of shared services and elimination of duplicative information technology (IT) investments, enhanced information through data standardization and system integration, and optimized service delivery through streamlining and normalization of business processes and mission operations.

DOE offers comments on the draft report and a status update for your consideration. Specifically, these comments are in reference to DOE's progress in measuring and reporting EA outcomes and benefits relative to elements of GAO's Enterprise Architecture Management Maturity Framework (EAMMF), which is documented in Table 1 and Table 10.

DOE Rating of "Not Satisfied" for Element #2 – A method and metrics have been established to measure enterprise architecture strategic mission value (outcomes and benefits): *Energy has not established metrics and a method for measuring enterprise architecture strategic mission value. Specifically, although the department's June 2012 draft Enterprise Modernization Roadmap includes potential enterprise architecture program metrics (e.g., cost savings through retiring legacy systems and cost avoidance by leveraging existing solutions over procuring new ones through the use of enterprise architecture), the metrics are still being defined and have yet to be finalized and approved. Regarding a methodology, the draft document states that appropriate processes will be developed once the metrics are developed and approved. According to the Chief Architect, the roadmap is expected to be completed and submitted to OMB by August 31, 2012.*

 Printed with soy ink on recycled paper

2

The DOE OCIO has established metrics and a methodology for measuring EA strategic mission value. Specifically, DOE has accomplished the following:

- Revised and submitted the Enterprise Modernization Roadmap (EMR) to the Office of Management and Budget (OMB) on August 31, 2012. The revised EMR centers on the new DOE IT Modernization Strategy presented to OMB on August 22, 2012 as part of the OMB PortfolioStat implementation. The Department is aggressively pursuing this new IT strategy to modernize DOE's infrastructure, deliver cost-effective commodity IT services, and improve the cybersecurity posture. A copy of the revised EMR is enclosed.

- Identified EA metrics across the Department in the following five categories:
  1) Cost Avoidance / Reduction;
  2) Compliance;
  3) Accuracy;
  4) Reuse; and
  5) Customer Satisfaction

- Established a methodology for metrics collection which focuses on the following three phases:
  1) Identification of Potential Metrics;
  2) Metrics Execution; and
  3) Project Closeout

- Implemented the DOE Enterprise Portfolio Management (EPM) system (Troux) that provides in-depth analysis and reporting to EA metrics.

- Additional EA Performance is included in section 2.7 of the EMR:

EA performance and value is demonstrated and measured at all levels within the DOE federated architecture model. At the OCIO level, specific metrics for the HQ-level EA program performance are still being defined, and will be aligned to the EA vision for value delivery at the Department. Potential metrics are listed below, with a description of the results demonstrated by each metric.

Potential DOE EA Metric	Results Demonstrated
Number of ARB Meeting Attendees	EA collaboration and communication
Number of Requests for Repository Data	EA use by program offices
Shared Services Opportunities Identified – Number	Identify potential applications which can be consolidated to a shared service
Shared Services Opportunities Identified – Cost Savings / Cost Avoidance	Document the cost savings (through retiring legacy applications) or cost avoidance (by leveraging existing solutions over procuring new ones) through use of the EA

Potential DOE EA Metrics

3

The metrics will be developed collaboratively, and once they are approved, appropriate processes will be developed to measure the EA results and outcomes. EA efforts at the Program/Staff Offices or lower levels have implemented a process to track EA results and outcomes through IT Modernization Initiatives (which includes IT Success Stories). Additionally, individual metrics may be identified and reported, based on the value proposition for the given effort. Examples of EA value delivery are given throughout the EMR, in the context of the specific initiative which achieved the success. A few examples are included in the table below.

Initiative	Performance Measure	Measurement
National Training and Education Resource (NTER)	Cost Savings	Estimated initial annual savings of over $175,000 compared to training system.
Data Center Consolidation	# Federal Data Centers Closed	Four data centers closed to-date (current); six additional closures planned (three in FY 2013 and three in FY 2015).
Public Key Infrastructure (PKI)	Reduction in legacy systems in use	50% reduction in systems to provide PKI services.

Examples of EA Value Delivery

DOE considers these accomplishments sufficient to justify a "Partially Satisfied" rating.

DOE Rating of "Not Satisfied" for Element #3 – Enterprise architecture outcomes and benefits are periodically measured and reported to the agency's enterprise architecture executive committee: *The department has yet to measure and report enterprise architecture outcomes and benefits.*

To measure EA outcomes and benefits DOE has accomplished the following:

- DOE is using the EA success stories collected from its program offices and labs to advertise and share as "Best Practices" and solutions that can be leveraged across the DOE Enterprise.
- Sharing DOE EA success stories at Architecture Review Board meetings (ARB), which is a sub-working group of the Information Technology Council (ITC). The ITC reports directly to the Information Management Governance Council (IMGC) which consists of the DOE Under Secretaries and the Chief Information Officer (CIO).
- Established an annual EA Awards Program to recognize individuals and DOE organizations who have achieved significant EA outcomes and benefits.
- Established an EA panel to formulate EA Success Charter.
- Established an EA knowledge base to enable EA methodologies.
- Established the DOE Architecture Review Board Wiki site (PowerPedia) to promote transparency and collaboration.

DOE considers these accomplishments sufficient to justify a "Partially Satisfied" rating.

4

DOE will continue to measure and mature the EA program based on the GAO's EAMMF (Version 2.0). Again, thank you for the opportunity to review the draft report. If you have any questions, please contact me on 202-586-4542 or at Rick.Lauderdale@hq.doe.gov.

Sincerely,

Walter R. Lauderdale
Chief Architect

Enclosure

Appendix XVIII: Comments from the Department of Housing and Urban Development

U.S. DEPARTMENT OF HOUSING AND URBAN DEVELOPMENT
WASHINGTON, DC 20410-4000

CHIEF INFORMATION OFFICER

SEP 0 7 2012

Ms. Valerie C. Melvin
Director, Information Management
 and Technology Resources Issues
U.S. Government Accountability Office
441 G Street, NW
Washington, DC 20548

Dear Ms. Melvin:

Thank you for the opportunity to comment on the Government Accountability Office (GAO) draft report entitled, *ORGANIZATIONAL TRANSFORMATION: Enterprise Architecture Value Needs to be Measured and Reported* (GAO-12-791).

The Department of Housing and Urban Development reviewed the draft report and respectfully submits the following comments:

Page 14, 2nd Bullet: "March 2012, the department had not yet finalized its architecture policy, as we had recommended".

Comment: HUD has a finalized Enterprise Architecture (EA) Policy that has been in place since April 10, 2002; we have updated that Policy, and we are currently in the process of achieving CIO concurrence and submitting it through the Departmental Clearance process for final approval.

Page 24 – 2nd Bullet: "HUD measured and completed its first report on outcomes, and highlighted......However, while the report was completed in December 2011, it had yet to be reviewed by agency executives. According to agency officials, the report is expected to be reviewed by the end of August 2012".

Comment: The CIO submitted HUD's EA Value Measurement Report for FY 2011 to the Customer Care Committee (CCC) for review on Wednesday, August 22. The CCC is comprised of HUD Executives within the overall HUD IT Governance Framework. See attached document for an overview of HUD's IT Governance structure.

Page 28 – after the second bullet, the paragraph starts by saying, "In addition, we recommend that the Secretaries of HHS and HUD ensure that enterprise architecture outcomes are periodically measured and reported to top agency officials".

Comment: HUD currently issues an Annual EA Value Measurement Plan and Report. The results of the measures in the plan will be documented in the report annually for the fiscal year.

2

Page 50 – 3rd element that is half full for HUD states, "Further, while the report was completed in December 2011… it has yet to be reviewed by agency executives".

Comment: HUD submitted the EA Value Measurement Report for FY 2011 to the Customer Care Committee (CCC) for review on Wednesday, August 22, 2012. The CCC is comprised of HUD Executives within the overall HUD IT Governance Framework. See the attached document for an overview of HUD's Governance structure.

HUD complied with the recommendation for executive action requiring that enterprise architecture outcomes are periodically measured and reported to top agency officials. With new EA value measurement and reporting guidelines to be issued by the Office of Management and Budget in late 2012, HUD will adhere to the new standards going forward.

If you have any questions, please contact Joyce M. Little, Director, Office of Investment Strategies Policy and Management, at (Joyce.M.Little@hud.gov), or 202-402-7404.

Sincerely,

Jerry E. Williams
Chief Information Officer

Enclosure

**Appendix XVIII: Comments from the
Department of Housing and Urban
Development**

HUD IT Governance Structure

HUD's IT governance structure is shown below. The structure empowers business areas to influence IT strategic priorities and ensure that all portfolio activities align with mission area needs. This process requires significant business area participation in the activities of the IT governance bodies illustrated below.

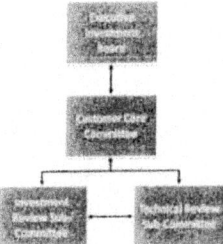

HUD's IT Governance Structure

The **Executive Investment Board** (EIB) comprises the Department's senior leaders. Its responsibilities include defining and implementing HUD's strategic direction, managing the Department's investment portfolio, and directly funding, overseeing, and reviewing complicated, costly, and highly visible projects.

The **Customer Care Committee** (CCC) comprises executives that manage HUD's IT investments and perform project oversight. The CCC's responsibilities include ensuring that investments and projects align with the Department's strategic plan, reviewing and submitting investment recommendations to the EIB, and coordinating with the Investment Review Sub-Committee and Technical Review Sub-Committee regarding investment and project management.

The **Investment Review Sub-Committee** (IRC) comprises business area personnel that focus on investment management oversight. Its responsibilities include reviewing all business cases to ensure their alignment with the Department's strategic goals, and providing guidance in the creation of OMB Exhibit 300s and Exhibit 53s. The IRC's members participate in the annual IT budget formulation process as representatives of their respective business areas.

The **Technical Review Sub-Committee** (TRC) comprises key personnel from within the OCIO. It focuses on project management oversight and technical architecture. Its responsibilities include ensuring that each segment architecture remains in alignment with the Department's strategic goals, monitoring all of HUD's IT projects, and providing analysis to the IRC, CCC, and EIB as needed. The TRC acts as a control gate in the PPM Life Cycle to ensure that necessary artifacts are produced and mandatory procedures are followed.

Appendix XIX: GAO Contact and Staff Acknowledgments

GAO Contact	Valerie C. Melvin at (202) 512-6304 or melvinv@gao.gov
Staff Acknowledgments	In addition to the contact named above, Neelaxi Lakhmani (Assistant Director), Mark Bird (Assistant Director), Virginia Chanley, Kelly Dodson, Cheryl Dottermusch, Michael Holland, James Houtz, Catherine Hurley, Stuart Kaufman, Lee McCracken, Tyler Mountjoy, Donald Sebers, Jennifer Stavros-Turner, and Merry Woo made key contributions to this report.

GAO's Mission	The Government Accountability Office, the audit, evaluation, and investigative arm of Congress, exists to support Congress in meeting its constitutional responsibilities and to help improve the performance and accountability of the federal government for the American people. GAO examines the use of public funds; evaluates federal programs and policies; and provides analyses, recommendations, and other assistance to help Congress make informed oversight, policy, and funding decisions. GAO's commitment to good government is reflected in its core values of accountability, integrity, and reliability.
Obtaining Copies of GAO Reports and Testimony	The fastest and easiest way to obtain copies of GAO documents at no cost is through GAO's website (http://www.gao.gov). Each weekday afternoon, GAO posts on its website newly released reports, testimony, and correspondence. To have GAO e-mail you a list of newly posted products, go to http://www.gao.gov and select "E-mail Updates."
Order by Phone	The price of each GAO publication reflects GAO's actual cost of production and distribution and depends on the number of pages in the publication and whether the publication is printed in color or black and white. Pricing and ordering information is posted on GAO's website, http://www.gao.gov/ordering.htm. Place orders by calling (202) 512-6000, toll free (866) 801-7077, or TDD (202) 512-2537. Orders may be paid for using American Express, Discover Card, MasterCard, Visa, check, or money order. Call for additional information.
Connect with GAO	Connect with GAO on Facebook, Flickr, Twitter, and YouTube. Subscribe to our RSS Feeds or E-mail Updates. Listen to our Podcasts. Visit GAO on the web at www.gao.gov.
To Report Fraud, Waste, and Abuse in Federal Programs	Contact: Website: http://www.gao.gov/fraudnet/fraudnet.htm E-mail: fraudnet@gao.gov Automated answering system: (800) 424-5454 or (202) 512-7470
Congressional Relations	Katherine Siggerud, Managing Director, siggerudk@gao.gov, (202) 512-4400, U.S. Government Accountability Office, 441 G Street NW, Room 7125, Washington, DC 20548
Public Affairs	Chuck Young, Managing Director, youngc1@gao.gov, (202) 512-4800 U.S. Government Accountability Office, 441 G Street NW, Room 7149 Washington, DC 20548

Please Print on Recycled Paper.

www.ingramcontent.com/pod-product-compliance
Lightning Source LLC
Chambersburg PA
CBHW081505170526
45166CB00008B/2562